Who greens the waves?

Who greens the waves?

Changing authority in the environmental governance of
shipping and offshore oil and gas production

Judith van Leeuwen

Environmental Policy Series – Volume 1

Wageningen Academic Publishers

ISBN: 978-90-8686-143-9
e-ISBN: 978-90-8686-696-0
DOI: 10.3920/978-90-8686-696-0

ISSN: 2210-3309

First published, 2010

© Wageningen Academic Publishers
The Netherlands, 2010

Preface

This book is the culmination of my PhD-research project. It explores the environmental governance of shipping and offshore oil and gas production from a historical perspective. During the years in which I have studied especially shipping and offshore oil and gas production, the degradation of the marine environment has entered the political agenda of the European Union and its member states, making it a current topic. I am very happy that this book contributes to these questions of the day and that this policy domain will remain the focus of my research for the years to come.

Theoretically, this PhD project embraces a topic that has captured my interest since the start of my academic career: how does international and European environmental policy come about? Why and how do states cooperate with each other to solve environmental issues internationally? When the literature took up 'governance' as a key concept in discussing European and international environmental policy making, a key question became: how do states cooperate with each other in an increasingly multi-actor and multi-level setting? And to what extent does international environmental governance still evolve around the state, or have other actors become more important?

When I started this PhD research project, my aims were to find empirical evidence for the shifts in governance from international governmental arrangements to innovative global governance arrangements in which market and societal stakeholders play an important role. A second aim was to find a way out of the conceptual confusion that exists about how to analyze the shift to innovative global governance arrangements and the question who is involved in what way in global policy processes. This book is the outcome of my attempt to achieve these aims.

I am thankful to many people for their direct and indirect contribution to this book. My colleagues at the Environmental Policy Group at Wageningen University have always supported me and ensured an enjoyable working environment. Special thanks goes to Prof. dr. ir. Arthur Mol and Dr. ir. Jan van Tatenhove, who have guided me during this research process, provided me with feedback and comments, and allowed me to shape the research project according to my own vision. I have interviewed many respondents to generate the information needed for this research project. Thank you for giving me some of your time and insights in the environmental governance of shipping and offshore oil and gas production. I want to thank all my friends who have maintained their interest in my work and the progress of my research project throughout the years. My family and family-in-law have always given me their unconditional support, thank you for that. Finally, André, you have always stood behind me and I enjoy the warmth and care you give me everyday.

Judith van Leeuwen

Wageningen, March 2010

Table of contents

Preface   7

Abbreviations   13

Chapter 1.
Introduction   15
  1.1 The marine environment as a 'new' policy domain   15
  1.2 Shifts in governance   16
  1.3 Changing authority in governance   17
  1.4 Research questions   18
  1.5 Outline of the thesis   19

Chapter 2.
Changing spheres of authority in governance   21
  2.1 Shifts in governance   21
    2.1.1 Multiple actors in governance   21
    2.1.2 Multiple levels in governance   22
    2.1.3 Multiple rules in governance   22
    2.1.4 Multiple steering mechanisms in governance   23
    2.1.5 Multiple spheres of authority in (global) governance   24
  2.2 The need for new theoretical concepts for governance   25
  2.3 New theoretical concepts for governance: a review   27
    2.3.1 Rule systems and sphere of authority   27
    2.3.2 Policy Arrangement Approach   30
  2.4 Spheres of authority in governance   32
    2.4.1 Redefining sphere of authority   32
    2.4.2 The organization of a sphere of authority   34
    2.4.3 The substance of a sphere of authority   36
    2.4.4 The results of a sphere of authority   37
  2.5 Changing authority in governance   38
    2.5.1 Renewal of an existing sphere of authority   38
    2.5.2 New spheres of authority in governance   40
  2.6 Political modernization in governance   41
  2.7 Conclusions: analysing the changing authority of the state in governance   42

Chapter 3.
Research methodology   45
  3.1 Operationalizing the research questions   45
  3.2 Case study and case selection   46
  3.3 Data collection   48
  3.4 Validity of the research   49

Chapter 4
Changing authority in the environmental governance of shipping   51
  4.1 The emerging global sphere of authority during the 1950s and 1960s   51
    4.1.1 The OILPOL Convention and the Intergovernmental Maritime
          Consultative Organization   51
    4.1.2 Freedom of the sea and flag states   53
    4.1.3 Private actors   55
    4.1.4 Compliance   56
  4.2 The authority of the state during the 1950s and 1960s   57
    4.2.1 Authority in developing steering mechanisms   57
    4.2.2 Authority in generating compliance   60
  4.3 The institutionalization of the port and coastal state during the 1970s and 1980s 62
    4.3.1 The MARPOL Convention and port state inspections   62
    4.3.2 The changing institutional structure of IMCO   63
    4.3.3 The United Nations Convention on the Law of the Sea   65
    4.3.4 Memoranda of Understanding on Port State Control   67
  4.4 The renewed global sphere of authority during the 1980s and early 1990s   68
    4.4.1 Ratification of the MARPOL Convention   68
    4.4.2 Flag, port and coastal states   69
    4.4.3 The level playing field discourse   71
    4.4.4 The threat of unilateral standards   73
  4.5 Further developments in the global sphere of authority during the 1980s and 1990s 74
    4.5.1 The entry into force of MARPOL Annexes   74
    4.5.2 Amending the Paris MoU on Port State Control   75
    4.5.3 New issues and Conventions   76
  4.6 The authority of the state during the 1980s and early 1990s   77
    4.6.1 Authority in developing steering mechanisms   78
    4.6.2 Authority in generating compliance   80
  4.7 The emergence of an European sphere of authority during the 1990s and 2000s   83
    4.7.1 European Directives and the European Maritime Safety Agency   83
    4.7.2 The EU sphere of authority and changes within IMO   86
  4.8 The EU sphere of authority and the renewed global sphere of authority during
      the 2000s   88
    4.8.1 Diverging European standards and IMO   88
    4.8.2 The continuing conventional IMO and the EU   91
    4.8.3 Technology-forcing standards within IMO   92

4.9 The authority of the state during the 2000s   92
    4.9.1 Authority in developing steering mechanisms after the EU policy change   93
    4.9.2 Authority in developing technology-forcing steering mechanisms   94
    4.9.3 Authority in generating compliance   96
4.10 Future outlook: initiatives outside IMO and EU   98

Chapter 5.
Changing authority in the environmental governance of offshore oil and gas production

                                                                                                101
5.1 Oil and gas production on the North Sea   101
5.2 The emerging regional sphere of authority during the 1970s and 1980s   101
    5.2.1 The Paris Convention and oil pollution   102
    5.2.2 National implementation: the UK, Norway and the Netherlands   107
5.3 The authority of the state during the 1970s and 1980s   112
    5.3.1 Authority in developing steering mechanisms   112
    5.3.2 Authority in generating compliance   114
5.4 The emerging national spheres of authority during the 1990s and 2000s   115
    5.4.1 United Kingdom   116
    5.4.2 Norway   118
    5.4.3 The Netherlands   122
5.5 The emergence of industrial steering mechanisms during the 1990s and 2000s   125
    5.5.1 Company environmental policies and environmental management systems   125
    5.5.2 Implications for the national spheres of authority   128
5.6 The renewed regional sphere of authority during the 1990s and 2000s   130
5.7 The authority of the state during the 1990s and 2000s   135
    5.7.1 Authority in developing steering mechanisms   135
    5.7.2 Authority in generating compliance   137
5.8 The emergence of an European Union sphere of authority during the 2000s   139
5.9 Renewed national and regional spheres of authority during the 2000s   141
    5.9.1 Renewed national spheres of authority   141
    5.9.2 Renewed regional sphere of authority   146
5.10 Authority of the state during 2000s   149
    5.10.1 Authority in developing steering mechanisms   149
    5.10.2 Authority in generating compliance   151
5.11 Future outlook   151

Chapter 6.
Comparing the environmental governance of shipping and offshore oil and gas production

                                                                                                153
6.1 Introduction   153
6.2 The nature of maritime activities   153

6.3 The evolution of the environmental governance of shipping and offshore oil and
    gas production   154
    6.3.1 Traditional spheres of authority   155
    6.3.2 Policy changes and the emergence of new spheres of authority   156
    6.3.3 Renewed spheres of authority   158
6.4 Shifts in governance   161
6.5 Changing authority of the state   163
    6.5.1 Traditional authority of the state   163
    6.5.2 Changing authority of the state in the development of steering mechanisms   166
    6.5.3 Changing authority of the state in generating compliance   169

Chapter 7.
Conclusions and reflection   171
7.1 Introduction   171
7.2 Shifts in governance: how innovative is environmental governance at sea?   172
7.3 Changes in governance and authority   174
    7.3.1 Conventional and new spheres of authority   174
    7.3.2 Changing governance   176
    7.3.3 Changing authority of the state   177
7.4 Reflection on research and the conceptual framework   178
7.5 Epilogue: future outlook for environmental governance at sea   180

References   183

Appendices   191

Summary   195

About the author   201

Abbreviations

ACOPS	Advisory Committee on the Prevention of Oil Pollution
CFc	chlorofluorocarbon
CHARM	Chemical Hazard Assessment and Risk Management model
DTI	Department of Trade and Industry (in the UK)
E & P Forum	Oil Industry International Exploration and Production Forum
EC	European Commission
EMSA	European Maritime Safety Agency
EOSCA	European Oilfield Speciality Chemicals Association
EU	European Union
FO-Industry	Facilitair Organisatie Industrie
FOEI	Friends of the Earth International
GOP	Working Group on Oil Pollution
HSE Regulations	Health, Safety and Environment in the Petroleum Activities Regulations
HMCS	Harmonized Mandatory Control Scheme
IMCO	Intergovernmental Maritime Consultative Organization
ICS	International Chamber of Shipping
Intertanko	International Association of Tanker Owners
IMO	International Maritime Organization
ISO	International Organization for Standardization
MARPOL Convention	International Convention for the Prevention of Pollution from Ships
MEPC	Marine Environmental Protection Committee
MNCs	Multinational corporations
MoU	Memorandum of Understanding
NGO	Non-Governmental Organization
NPD	Norwegian Petroleum Directorate
NOGEPA	Netherlands Oil and Gas Exploration and Production Association
NSMC(s)	North Sea Ministerial Conference(s)
OGITF	Oil and Gas Industry Task Force
OGP	International Association of Oil & Gas producers
OIC	Offshore Industry Committee
OILPOL Convention	International Convention for the Prevention of Pollution of the Sea by Oil
OLF	Norwegian Oil Industry Association
OPEC	Organization of the Petroleum Exporting Countries
OSPAR Convention	Convention for the Protection of the Marine Environment of the North-East Atlantic
Oslo Convention	Oslo Convention for the Prevention of Marine Pollution by Dumping from Ships and Aircraft
Paris Convention	Paris Convention on the Prevention of Marine Pollution from Land-based Sources

ppm	parts per million
REACH	Registration, Evaluation, Authorisation and Restriction of Chemicals
SEBA	Working Group on Sea-Based Activities
SFT	Norwegian Pollution Control Authority
TBT	tributyl tin
UK	United Kingdom
UKOOA	United Kingdom Offshore Operators Association
UN	United Nations
UNCLOS	United Nations Convention on the Law of the Sea
US	United States of America

Chapter 1.
Introduction

1.1 The marine environment as a 'new' policy domain

The marine environment is under increasing pressure of maritime activities. Those who think the sea is empty are in for a surprise. Numerous activities take place at sea, with sometimes adverse effects for the marine environment. In fact, Halpern and his colleagues (2008) give evidence that no area is unaffected by human activities and that the Eastern Caribbean, the North Sea and Japanese waters are the most affected regions. These regions are affected by activities like shipping, gas and oil production, sand extraction, production of energy in wind mill farms, fisheries, aquaculture and tourism. These maritime activities are associated with environmental effects such as oil spills, chemical discharges, depletion of natural resources like fish, air pollution, waste generation and noise pollution.

What is more, the scale of maritime activities is increasing. The past growth in shipping is expected to continue in the (long-term) future because of increases in population and consumption in especially Asia. Wind mill farms are a new phenomenon and long-term experiences with building and maintaining them in a sustainable manner is lacking. The demand for fish as part of a healthy diet is increasing as well, which increases the pressure on existing fish stocks, which are already overfished, and enlarges the demand for aquaculture.

The intensified use of the sea has, at least in Europe, led to new governance initiatives to deal with the resulting environmental effects and risks for the marine environment. For example, the European Commission proposed a European Integrated Maritime Policy and the Marine Strategy Framework Directive has already been adopted to protect the marine environment of European seas. The appointment of Marine Protected Areas in the North Sea is also underway.

Yet, those involved in policy-making for the protection of the marine environment face a number of challenges. These challenges make the protection of the marine environment a breeding ground for governance initiatives. First, the increase in the number and intensity of maritime activities also increases the stakes involved in developing environmental policies for the sea. The increase in stakeholders involved, with each their own interests, makes developing policies a challenging matter.

Second, pollution and activities at sea are not always occurring in the territory of a state. The further one goes away from the coastline, the less a state can rely on traditional sovereignty. Different jurisdictional arrangements exist for the areas till 12 nautical miles (nm), 200 nm (the exclusive economic zone), the continental shelf and the high seas. With each boundary the autonomy of the state to regulate decreases, with the high sea being a common where individual states have no autonomy to regulate at all. The variety of stakes involved and the fact that the states do not have the autonomy to develop policies means that one can expect innovative governance arrangements that regulate the environmental impact of the various maritime activities taking place at sea.

Third, pollution at sea is by definition transboundary. Pollution enters the marine environment and spreads with wind, currents and tides. A good example is the incident where a container with 29,000 bath toys fell off a container ship in the Pacific Ocean in 1992. These ducks have drifted all over the world and scientists have followed these ducks to increase knowledge on ocean currents. Moreover, one sea is generally shared by a number of coastal states, meaning that pollution often affects a number of states at the same time. For example, the North Sea has 8 bordering states: Norway, Sweden, Denmark, Germany, the Netherlands, Belgium, France and the United Kingdom. Protecting the North Sea therefore demands cooperation between these states. Transnational environmental policies and cooperation are thus essential in preventing and minimizing the pollution of maritime activities. The increased stakes involved, the issue of decreasing sovereignty and the need for transnational cooperation mean that there is a need for different types of governance arrangements to regulate actors' behaviour at sea and the environmental risks and consequences of it.

Interestingly, transnational cooperation in the environmental governance of maritime activities is – in fact – not that new. Especially the environmental effects of shipping, oil and gas production in the North Sea and fisheries in Europe are already regulated since respectively the 1950s, the late 1970s and the early 1980s. The long history of environmental governance at sea and the special characteristics of the sea make the protection of the marine environment an interesting policy domain for a study into how environmental governance practices have changed under influence of globalization in the last decades. This policy domain allows for an analysis into how conventional governance arrangements are challenged and transformed by more recent innovative governance arrangements. Moreover, despite a long history in environmental governance for the sea, this policy domain has not received much scrutiny by social scientists; another reason to turn our attention to the marine environment as a policy domain for a study into shifts in governance.

This thesis therefore aims to use the 40 years of experience with environmental governance on the North Sea to better understand a number of the essential theoretical issues discussed in governance literature. In doing so, the thesis will focus on shipping and offshore oil and gas production in particular as these are the maritime activities that have the longest history in being subject to specific governance arrangements that regulate the environmental impact of their operations.

1.2 Shifts in governance

A much discussed subject within social sciences literature is that (environmental) governance practices are changing as a result of globalization processes. Although globalization affects all aspects of society (see for example Held & McGrew, 2002; Held, McGrew, Goldblatt & Perraton, 1999; Keohane, 2002; Prakash & Hart, 1999; Scholte, 2000; Waters, 2001), this thesis focuses on the transformations in the environmental and political realm as part and result of globalization processes because it is the global nature of environmental issues itself that is radically changing conventional ideas of policy making, governments and governance (Bruyninckx, Spaargaren & Mol, 2006). This thesis aims to contribute to the theoretical debate on the changing political realm of society by studying changing governance practices in more detail.

The concept of governance has become very popular within the social sciences, politics and society. The concept is, however, also clouded by conceptual confusion (see among others Dingwerth & Pattberg, 2006; Kjaer, 2004; Treib, Bähr & Falkner, 2005, 2007; Van Kersbergen &

Van Waarden, 2004). Although some authors define governance as being opposed to government, this thesis regards government as part of governance. Governance is defined in broad terms as 'steering of society', hence including conventional governmental practices.

But even within such a broad notion of governance, there is, very generally speaking, a shift from government to governance. Underlying this shift are shifts in the locus and focus of governance. The locus of governance is changing, because of shifts to multiple actors and levels. The political system is increasingly characterized by multi-actor and multi-level features, because there has been a proliferation of international institutions that facilitate the development of international policies. Numerous non-governmental organizations (NGOs) have emerged on the local, national and global level, sometimes bypassing the state in their attempts to influence policy making and implementation. Multinational corporations (MNCs) have become political actors and individual corporations are joining their forces through the establishment of associations at the transnational level.

At the same time, transformations in governance show a shift in the focus of governance, because rules of the games that shape governance practices are changing and need to be renegotiated. The conventional democratic rules that arrange who is accountable and what is legitimate are changing and are renegotiated during the establishment of new governance arrangements. Moreover, the output of governance practices in terms of type of steering mechanisms is also shifting from formal, national laws to voluntary agreements or norms.

Yet, despite this diversification of actors, levels, rules and steering mechanisms involved in (environmental) governance practices, the conventional way of governing is not disappearing or completely replaced by new practices. Both Beck (2005) and Rosenau (1997, 2000, 2002) argue that a new world of world politics has emerged next to the already existing world in which world politics is based on a system of nation-states and treaty formation. The old world politics is characterized by actors applying the already existing rules for decision making and implementation. In international politics, this conventional way of governing manifests itself through the establishment of treaties. In contradiction, the new world consists of a multicentric system of diverse types of collectivities and has emerged as a rival source of authority that sometimes cooperates with, often competes with, and endlessly interacts with the state-centric system (Rosenau, 2002). In other words, multiple sources of authority have emerged, consisting of a combination between conventional governance practices and new ones.

The aim of this thesis is to report on a research into the different sources of authority in the environmental governance of shipping and offshore oil and gas production. This thesis will explore how the environmental effects of the two maritime activities were governed traditionally, it will analyse which shifts in governance have affected the environmental governance of these maritime activities and it will look into how this has led to a multi-centric system of sources of authority in the environmental governance of shipping and offshore oil and gas production.

1.3 Changing authority in governance

The changing character of (environmental) governance has implications for the actors' authority in governance practices. There is a current debate on the extent in which the authority of the state has eroded as a consequence of shifts in governance. At the onset of environmental policy development, the nation state became 'the principle vehicle for the establishment of collective social goals and their attainment' (Waters, 2001, p. 13). To attain national environmental goals, states started to build relations with other states, because of the transboundary nature of many environmental issues. Yet, the distinction between domestic and foreign affairs remained clear-cut (Held *et al.*, 1999). However, the last decades this system of inter-state relations expanded to a system of relations between all kinds of actors and between all kinds of levels resulting in 'major shifts in the location of authority and the site of control mechanisms' (Rosenau, 1997, p. 153). As a result of public-private and private initiatives in environmental governance both at the global level, as well as at the regional, national and local levels, the position and role of the nation state is challenged and transformed.

Yet, the notion of authority moving away from the state as if it is a zero-sum game is better replaced by a more nuanced view about the continued importance of the state, because despite the shifts in governance 'it should be clear that states are not about to walk off the world stage' (Rosenau, 1997, p. 347). However, 'nation states should be seen no longer as 'governing' powers, able to impose outcomes on all dimension of policy within an given territory by their own authority, but as loci from which forms of governance can be proposed, legitimated and monitored' (Hirst & Thompson, 1996, p. 190). Thus, while the nation-state continues to have authority in environmental governance practices, its conventional authority is challenged and transformed because of shifts in governance. This thesis aims to increase understanding about how the state continues to play a substantial, be it altered, role in governance practices as a result of the emergence of multiple sources of authority. In doing so, it goes beyond the debate on whether the authority of the state is eroding or not and aims to show that the authority of the state is changing.

1.4 Research questions

The aim of this thesis is to study how the authority of the state is changing as a result of shifts in governance. This also entails the development of a conceptual framework to analyse changing governance practices and what this means for the authority of actors, in particular the state, because this is currently missing in social sciences. Assuming that the state is at the heart of traditional governance practices within the environmental governance of shipping and of offshore oil and gas production, this thesis will, analyse how governance practices have opened up to levels, actors, rules and steering mechanisms beyond the state. This in turn challenges the monopolistic position and authority of the state and will result in changes in authority in the environmental governance of shipping and of offshore oil and gas production.

In sum, this thesis seeks to generate new insights for the governance literature by gaining an increased understanding of shifts in governance and the associated changes in authority of the state that are currently occurring within the environmental governance of shipping and of offshore oil and gas production. The objective of this research is therefore *to analyse how shifts in governance affect and change the authority of the state in the environmental governance of shipping and of offshore oil and gas production in order to generate new insights for the governance debate.* In order to achieve this objective, the following *preliminary* research questions are asked at the start of this thesis:

- Which shifts in governance have taken place in the environmental governance of shipping and of offshore oil and gas production?
- How has the authority of the state in the environmental governance of shipping and of offshore oil and gas production changed under influence of shifts in governance?
- What are the insights gained from this research for the governance debate?

These research questions are preliminary research questions, because they will be operationalized and further refined after the conceptual framework is developed in Chapter 2. As mentioned, governance literature debates shifts in governance and the changing authority of the state, but lacks a theory or conceptual tools to do so. That is why these conceptual tools are developed first, before these research questions can be further operationalized to guide the rest of this thesis.

1.5 Outline of the thesis

The next chapter of this thesis, Chapter 2, will start with a further elaboration on the shifts in governance that are changing the political realm of society. The need for new theoretical concepts to analyse shifts in governance and the changing authority of actors, in particular the state, is the second issue addressed in this chapter. The majority of this chapter, however, deals with the development of a conceptual framework that is needed for this research and that forms an important part of the contribution of this thesis to the governance debate.

Chapter 3 will deal with the methodological decisions that have been made for the research underlying this thesis. The first one is the operationalization of the research questions based on the conceptual framework developed in Chapter 2. After that, the research design that is used in this thesis, i.e. the case study, and methods of data collection will be discussed. The validity of the research will also be reflected upon.

The empirical results of the research will be discussed in Chapters 4 and 5. Using the concepts developed in Chapter 2, these chapters show how the environmental governance of shipping and offshore oil and gas came about during the 1950s and late 1970s respectively. It will be analysed what the authority of the state was during these first years. Yet, the majority of these chapters analyse in detail how the environmental governance of the two maritime activities changed over the years. The implications of these changes for the authority of the state will be analysed as well.

In Chapter 6, the empirical results discussed in Chapters 4 and 5 will be compared. By comparing the way in which the environmental impacts of shipping and offshore oil and gas production were governed initially, how the environmental governance of both activities changed over the years and what the implications have been for the authority of the state, this chapter will already answer some of the research questions posed in Chapter 2. For example, conclusions will be drawn about the extent to which the environmental governance of shipping and of offshore oil and gas production have been affected by shifts in governance.

The rest of the research questions will be answered in the final chapter of this thesis, Chapter 7. This chapter lies down the conclusions of this thesis and will give a reflection on research done and the conceptual framework used in this thesis.

Chapter 2.
Changing spheres of authority in governance

2.1 Shifts in governance

This chapter will introduce the theoretical perspective and conceptual framework through which changes in the environmental governance of shipping and of offshore oil and gas production will be analysed. Moreover, it provides concepts to analyse the changing authority of the state in governance. The starting point for the conceptual framework is the four shifts in governance that have been observed over the last decades and which have caused the emergence of new spheres of authority in governance.

The concept of governance has become tremendously popular among social scientists. However, this popularity is accompanied by a very diverse use of the term governance (see among others Dingwerth & Pattberg, 2006; Kjaer, 2004; Treib *et al.*, 2005, 2007; Van Kersbergen & Van Waarden, 2004). Some authors use governance to refer to a new way of developing steering mechanisms through network-structured political processes (e.g. Gilpin, 2002; Rhodes, 2000). They speak of a shift from government to governance; from hierarchy to networks. Other authors have a broader notion of governance and view it as all steering of society (e.g. Pierre & Peters, 2000; Rosenau, 1997).

All these authors have in common, however, that they observe and theorize about recent shifts in governance practices. Based on Van Kersbergen & Van Waarden (2004) and on Treib, Bähr and Falkner (2005, 2007), I argue that four shifts characterize governance practices, namely that respectively multiple actors, multiple levels, multiple rules and multiple steering mechanisms have become features of governance practices. This section analyses these four shifts in governance and ends with the claim that these shifts have led to the coming about of new sites of politics, authority and steering.

2.1.1 Multiple actors in governance

The first shift that the concept of (global) governance builds upon is the fact that political processes increasingly involve multiple actors: 'by global governance is meant not only the formal institutions and organizations through which the rules and norms governing world order are (or are not) made and sustained – the institutions of state, intergovernmental cooperation and so on – but also all those organizations and pressure groups – from MNCs, transnational social movements to the plethora of non-governmental organizations – which pursue goals and objectives which have a bearing on transnational rule and authority systems' (Held *et al.*, 1999, p. 50).

There are a number of ways in which these actors challenge the traditional international politics within and between states. For example, while traditionally the access for NGOs to transnational or international politics was via the state, they now sit at the table directly. Besides this growth of access points, both NGOs and multinational corporations have gained much more expertise, while the state seems to lose this expertise. Moreover, NGOs and MNCs are increasingly developing steering mechanisms without state involvement. This has enhanced the position of NGOs and MNCs within political processes. In sum, three of the main challenges for the state are (1) that it has lost expertise to private actors, (2) that it has lost its role as a gate-keeper for participation of private actors in international policy making and (3) that it has lost the monopoly to develop and implement steering mechanisms.

However, the increasing interactions between state actors and non-state actors can also help in developing policies and implementing them effectively. The increasing participation of non-state actors can ensure feasible and effective policies, because these actors can help to determine which goals are feasible or not; to which kind of policies they will confine themselves and which not. In doing so, the interaction with non-state actors can help the state in pursuing its interests. Whether this increasing number of actors participating in political processes facilitates or obstructs developing and implementing effective policies depends on the issue involved and the political process itself.

2.1.2 Multiple levels in governance

The increase in participants in governance goes hand in hand with the increasingly multi-level characteristic of political processes. We have seen that especially the number of international actors has grown in the last decades. However, also the issues and activities that demand governance are increasing in scale, which have resulted in more demand for global governance. The global economy, global environmental problems and human rights issues exemplify this. But there is not only an 'upward' trend to the international level, but also a 'downward' trend to the sub-national level. Policy competences are decentralized to the sub-national levels. For example, Hooghe and Marks (Hooghe & Marks, 2001; Marks, Hooghe & Blank, 1996) emphasize that within the European Union (EU) sub-national actors participate in policy making and engage in interactions with EU institutions directly.

The consequences of this shift for the state are, just like the consequences of the increasing number of actors, twofold; it can be both an advantage and a disadvantage to the state to participate in politics at various governance levels. It is often argued that the state regains control over issues that cross borders, because it cooperates with other state and/or non-state actors to tackle these issues. Some decades ago – when international policy making was not common practice – the state had lost grip on issues and problems that were global, because an individual state was not able to resolve the entire issue on its own. Yet, cooperating with other state and non-state actors often means that one has to be satisfied with a political compromise. In other words, a state can have less leeway to handle the issue in a way that matches its own interests, traditions and culture.

2.1.3 Multiple rules in governance

The diversification of rules that shape policy practices is the third shift within governance. The existing formal rules for democratic decision making and implementation have become inadequate because of among others the participation of new actors and levels in governance. These new actors and levels have broken up existing political processes and triggered the emergence of new sites and processes of politics. In these new sites and processes of politics, the existing rules are inadequate. There is thus institutional ambiguity (Hajer, 2003) and a need to formulate new rules of the game. These new rules of the game are negotiated simultaneously with the development of new policies. These new rules are often informal rules. Yet, when these new political processes become stable, the informal rules can develop into new formal rules.

The fact that existing rules are increasingly redefined and renewed in governance has consequences for the authority of the state. While the conventional democratic rules put the state at the epicentre of political processes and partly determined the authority of the state in these practices, the state is now subject to the new and informal rules themselves. This means that the conventional authority of the state itself is questioned during the definition of the new rules. Moreover, if the new rules, in the most extreme case, abandon or strongly limit the authority of the state in the new policy practices, the state has lost much of its original steering capacity.

However, on the other hand, the state is often involved in defining these rules. The state can thus take part in building its own position in the new political process. Another possibility is that the new actors are redefining the rules to create a stronger steering authority for the state. Subsequently, new and informal rules are not always a threat to the state, but can also enhance the authority of the state in governance.

2.1.4 Multiple steering mechanisms in governance

The fourth shift that is captured by the concept of governance lies in the variety of steering mechanisms, i.e. in the types of law and regulation that are used to steer certain actors and activities. Traditionally, the focus was on legally binding law that sets specific rules, goals and standards. However, new types of law and regulations are emerging, the so-called soft law and procedural regulation (Treib *et al.*, 2005, 2007).

Legally binding law is associated with different enforcement mechanisms to reach compliance than for example soft law. Legally binding law can rely on a whole set of formal procedures and mechanisms, like sanctions or jurisdiction, while soft law relies more on coordination, political will and pressure to reach compliance. In addition, the development of legally binding law with its rules and standards is much more based on existing procedures than soft law and procedural regulation. Especially procedural regulations arrange new procedures to raise environmental awareness and to include private actors (Treib *et al.*, 2005, 2007). These multiple steering mechanisms, such as hard law, norms, protocols, best practices, etc., can co-exist next to each other and there is not necessarily one that is dominant.

One can argue that the newly emerging steering mechanisms have different implications: they confront states with new procedures, they give much more room for private actors to participate in both decision making and implementation processes and they make pressure politics and the development of political will more important. Some authors therefore argue that the state seems to lose some of its power, because the state is no longer able to develop and enforce regulation through formal rules and standards on its own.

However, others argue that the shift to soft and procedural regulations enhances the authority of the state. The fact that states loosen their grip on the development and implementation of rules and goals pays itself back through the achievement of goals and standards that would otherwise not have been achieved. The specific consequences of this shift for the authority of the state in governance depend of course on the levels and laws or policies that are involved in governing an issue, activity or problem. It also depends on the mixture of conventional and new types of laws and regulations that are used within steering processes.

2.1.5 Multiple spheres of authority in (global) governance

The four shifts in actors, levels, rules and steering mechanisms together have led to the emergence of new and unconventional 'spheres of authority' that challenge, transform, but also complement the continuing conventional spheres of authority. According to Rosenau (2002), the term 'sphere of authority' refers to collectivities that have the ability to generate compliance on the part of those actors, groups, or persons to whom the governance practices are directed. Traditionally, authority was mostly in hands of state actors and the sphere of authority was hierarchical in nature, because the state governed through conventional formal and standardized interactions with only limited room for informal politics and non-state actors.

The four shifts towards multiple actors, levels, rules and steering mechanisms within political processes have thus given rise to new loci of authority and new spaces of politics besides the conventional nation-state system of politics; the shifts have triggered the building of networks of actors and interactions that go beyond the conventional boundaries of politics. In the words of Beck (1994, p. 18): 'the political has broken open and has erupted beyond formal responsibilities and hierarchies'. The result has been the blurring of boundaries between the state, market and civil society. While politics has entered the market and civil society, the market and civil society are increasingly involved in state business and political processes. It is in fact in the blurring of these boundaries that new spheres of authority emerge.

Both Beck (2005) and Rosenau (1997, 2000, 2002) explain the effect of the emergence of new sites of politics for *global* governance. They argue that a 'new world' of global politics has emerged next to the already existing world of global politics that is based on a system of nation-states and treaty formation. The old world of global politics, which continues to play a role in global governance, is characterized by actors applying the conventional rules for decision making and implementation, i.e. politics used to operate 'within the rule system of industrial and welfare state society in the nation-state' (Beck, 1994, p. 35). In other words, conventional politics was rule-directed (Beck, 1994).

In contradiction, the new world of global politics consists of a multicentric system of diverse types of collectivities and has emerged as a rival source of authority that sometimes cooperates, often competes, and endlessly interacts with the state-centric system (Rosenau, 2002). The new world of global politics is characterized by a logic of rule change in which rule-altering politics is resulting in a redefinition of the role of actors, new resources for actors, unfamiliar rules and new conflicts (Beck, 1994, 2005). This rule-altering politics takes place in an institutional void because it lacks generally accepted rules and norms (Hajer, 2003). This means that within this new world of global politics 'actors not only deliberate to get to favourable solutions for particular problems, but while deliberating they also negotiate new institutional rules' (Hajer, 2003, p. 175-176).

Thus, in conventional politics, rules are relatively formal and the process in which especially state actors are engaged in revolves around the definition of the problem at hand and how to solve the problem. In the conventional world of global politics, the conventional way of politics manifests itself through the establishment of treaties. The policy processes taking place in the new sites of politics evolve in a more chaotic and informal way because there are unconventional actors and governance levels involved and each actor has to (re)define and construct its own role. Moreover, actors have to construct new rules and are engaged in politics at the same time.

The effect of the emergence of politics in new sites and new spheres of authority on the steering capabilities of the state can be twofold; sometimes this shift can enhance the capacity of the state to govern issues or actors, sometimes the shift undermines the state. Many academics have been discussing the challenge that these new spaces pose for the state whose governing capabilities are traditionally based on the nation-state political system. It is argued that even though the role and authority of the state cannot be neglected in governance, the new spaces of politics and authority mean that states are less able to rely on the effectiveness of their authority and steering mechanisms. States still have a prominent place in governance, but they have lost some of their earlier dominance in the governance system and have less ability to evoke compliance and govern effectively, because of the growing relevance and potential of transnational and subnational systems of rule (Rosenau, 1997). Pierre (2000) joins Rosenau by observing that the conventional power bases of the state seem to be losing their former strength, but that there are alternative strategies through which the state is pursuing collective interest. Beck (2005) is also optimistic about the future of the state, because he believes that states have the capacity to act and transform itself to adapt to the current shifts in (global) governance.

2.2 The need for new theoretical concepts for governance

The emergence of new spaces of politics and new loci of authority challenge the conventional theoretical approaches towards both international and domestic politics, i.e. the sub-disciplines of International Relations and Public Administration. These conventional theoretical approaches generally suffer from methodological nationalism (Beck, 2005), which make them ill-equipped for inquiries into the political processes of today. This methodological nationalism refers to several fundamental features of conventional theories that cause the misfit between current political practices and conventional theories.

First, the conventional theories focus solely on the nation-state, because the assumption of these theories is that political processes always evolve around the nation-state. However, 'understanding is no longer served by clinging to the notion that states and national governments are the essential underpinnings of the world's organization' (Rosenau, 1999, p.287). It is generally acknowledged that the spheres of authority within (global) governance range from solely based on private actors to solely based on public actors with in between spheres of authority that are based on partnerships between public and private actors. This implies that we need new tools and theories that allow that political collectivities are not only formed by the state, but by other actors as well. Moreover, a research that is focused on the changing authority of the state – like this one – needs concepts that allow for the erosion and dispersion of state authority and the increasing relevance of other actors (Beck, 2005; Hajer, 2003; Rosenau, 1997, 2000).

The second misfit that is directly related with this state-centredness of the conventional theories is that conventional theories always have relied on an analytical boundary between domestic and foreign politics (Beck, 2005). The analysis and explanation of domestic politics is the domain of public administration scientists, while foreign politics has been subject to analysis by international relations theory. Both branches of theory have developed their own concepts, assumptions, and methodologies to analyse their political domain. However, the strong linkage between various governance levels through actors, rules and steering mechanisms is central to the governance concept. This implies that the transnational political arenas can no longer be analysed separately from the domestic level. This multi-level feature of governance is therefore fundamentally different from the two-level game (Putnam, 1988) between international and domestic politics that is associated with conventional International Relation theory. A new governance theory should thus allow for a multi-level perspective on political processes.

Moreover, the state-centredness of the conventional theories is associated with a focus on formal procedures, established habits and well-shared norms. But these established procedures, habits and norms are not adequate for the new spaces of politics. That is why both Beck (1994, 2005) and Hajer (2003) emphasize the new and informal rules that guide the political processes in the new spaces of politics. This thus means that a theoretical framework intended to analyse governance should incorporate the rule-altering nature of political processes in the new sphere of authority.

Finally, the state-centred conventional theories also share a focus on command-and-control as the expected output from state-based political processes. Steering is associated with formal laws (national laws or, multilateral treaties) and formal sanctions (at the national level), instead of with the plethora of steering mechanisms that are used in governance today. A new theory on governance should thus be able to deal with different types of steering mechanisms that come out of political processes.

This bias of conventional theories to politics within a nation state system is among others captured by the concepts that are used within theories of International Relations and Public Administration, such as nation-state, society, domestic politics, regulation, territory, etc. These concepts specifically refer to the nation-state system of politics. However, there is a need for new theoretical concepts that help us to grasp the new spaces of politics, but that at the same time allow for the still existing politics of the nation-state. Moreover, these concepts should not have the connotations and assumptions that go with the nation-state-based theories.

There are many scientists that are struggling with theorizing (global) governance and that are searching for new concepts, just like me, to go beyond conventional theories that focus on the territorial and nation-state based political system. However, I argue that some of these scientists have chosen concepts or approaches that suffer from the same sort of bias and connotations as the conventional theories and concepts. For example, a classification in terms of 'old' and 'new' modes of governance or of government versus governance also suffers from these specific connotations: old governance and government refer to steering based on nation-states, while (new) governance refers to steering through network-like structures. I argue that this classification between old and new governance and between government and governance is of little analytical value, because some modes of governance may have been relatively new in some empirical contexts, but may turn out to be long-established practice in other areas or states (Treib *et al.*, 2005, 2007). It should thus be avoided to use the concept governance in a way that limits it to network-like political processes only.

In this thesis, the analysis of governance is therefore approached from a broad perspective, i.e. governance is viewed as 'the steering of society'. The first reason to have a broad notion of governance as the starting point for this thesis is that it allows you to view, analyse and compare existing governance arrangements with newly emerging governance arrangement. As argued above, the fact that new sites of politics have emerged, does not mean that the conventional political processes have ceased. Thus to analyse governance and changes in governance both the conventional world of politics and the new world of politics should be researched.

A second reason is that governance and the concepts with which governance practices are investigated should be devoid of the a-priori assumption that existing governance arrangements are state based, hierarchical in nature and based on command-and-control regulation and that the new ones are not state based, horizontal in nature and based on soft regulation. In other words, the new concepts for governance should use analytical categories and concepts 'that describe the typical properties of governing modes rather than labels that refer to the point of time of their occurrence in specific empirical contexts' (Treib *et al.*, 2007, p. 2).

2.3 New theoretical concepts for governance: a review

Now that the need for new concepts is established, the question is which concepts or theoretical approaches can be used as a starting point for developing a conceptual framework for this thesis? As argued above, these concepts should allow for the analysis of both conventional and new forms of governance by using properties to define governance practices instead of labels. Moreover, since the objective of this research is to analyse the changing authority of the state, the conceptual framework should also deal with the way in which the authority of actors is embedded within governance practices. In this section, I will explore the extent to which the approach to global governance of Rosenau and the policy arrangement approach of Van Tatenhove and colleagues are useful starting points for this thesis.

2.3.1 Rule systems and sphere of authority

Rosenau (1997, p. 145) conceives governance 'as spheres of authority at all levels of human activity – from the household to the demanding public to the international organization – that amount to systems of rule in which goals are pursued through the exercise of control'. In this conception of global governance, Rosenau puts a lot of emphasis on control and steering (Dingwerth & Pattberg, 2006). He wants to do away with the notions of 'command and control' and command mechanisms, because they imply a hierarchy in the development of goals, policies and the achievement of those goals. Instead, he opts for using the term control or steering mechanisms, because these 'highlight the purposeful nature of governance without presuming the presence of hierarchy' (Rosenau, 1997, p. 146; 2006, p. 122).

Control is a relational phenomenon between the controllers and the controllees. The controllers seek to change the behaviour of other actors, the controllees. The controllees either resist or comply with the wish or requirements of the controllers. Rosenau (1997, 2006) has used the concept 'system of rule' to refer to this interaction between controllers and controllees. However, this interaction between the control efforts of the controllers and the compliance by the controllees has to have a certain level of regularity before it can be called a rule system (Rosenau, 1997, 2006). A system of rule thus refers to a more or less stable relationship between the controller and controllee.

In other occasions, however, Rosenau seems to regard rule systems not as a relational phenomenon, but as the steering mechanisms through which the relation between controllers and controllees is established. For example, he has argued that 'both government and governance consist of rule systems, *of steering mechanisms* through which authority is exercised [...]' [italics added] (Rosenau, 2002, p. 72). Later, he defines governance as 'sustained by *rule systems that serve as steering mechanisms* through which leaders and collectivities frame and move toward their goals' [italics added] (Rosenau, 2003a, p. 393). Or 'the core of governance involves rule systems in which steering mechanisms are employed to frame and implement goals that move communities in the directions they wish to go or that enable them to maintain the institutions and policies they wish to maintain' (Rosenau, 2003b, p. 13).

Just like Dingwerth and Pattberg (2006, p. 5), who argue that 'systems of rule exist where a number of mechanisms are in place which relate to each other and which regulate or have an impact on the norms, expectations, and the behaviour of the relevant actors within the regulated area', I will conceive rule systems as referring to the steering mechanisms that shape the controller-controllee relationship.

The main reason for this is that next to system of rule, there is another concept that is central to Rosenau's conception of global governance: sphere of authority. This concept also refers to a relational phenomenon, because authority is by no means subscribed to a positional quality or possession of actors, but viewed as being relational. The existence of authority can only be observed when authority is both exercised and complied with, because 'authority at work in a situation is to be found in the extent to which those toward whom such efforts are directed comply with the directives' (Rosenau, 2003a, p. 274). In other words, authority – just like control – refers to a relationship between the controllers and the controllees, in which the controllers try to reach certain goals or set certain rules, while the controllees are resisting or complying with these goals and rules.

The concept sphere of authority does not only refer to this authority-compliance relationship between actors, but also to *those collectivities* that have the authority to evoke compliance from those who are ruled (Rosenau, 2003a, 2006). And because a sphere of authority refers to the collectivities that have authority, I assume that rule systems should refer to the steering mechanisms that either support or generate authority. Together, the collectivity and its steering mechanisms bring about this authority-compliance relationship between the collectivity and the actors that are ruled.

In sum, to freely translate the conception of global governance that Rosenau has, global governance is about those collectivities (spheres of authority) and those steering mechanisms (systems of rule) that generate compliance from the target group. Yet, the three concepts of collectivity, steering mechanism and compliance are not further defined by Rosenau. With regard to the term collectivities Rosenau (2006, p. 29) argues that next to the state there is 'a wide range of other types of collectivities, from corporations to professional associations, from neighbourhoods to epistemic communities, from nongovernmental organizations to social movements, from professional societies to truth commissions, from Davos elites to trade unions, from subnational governments to transnational advocacy groups, from networks of the like minded to diaspora, from gated communities to vigilante gangs, from credit-rating agencies to strategic partnerships, from issue regimes to markets, and so on [...]'. A definition of collectivity would thus be something like 'an arrangement of actors'.

Rosenau also talks in more detail about steering mechanisms (Rosenau, 1997, 2006). He argues that they can be sponsored by states, by actors other than the state, or jointly by state and non-state actors. He also states that they can be placed on a continuum from nascent to institutionalized control mechanisms. Yet, Rosenau never explains what he means by steering mechanisms other than that they are a means through which control or authority is exercised. A definition of steering mechanism could thus be 'a set of instruments which regulates the behaviour of a targeted (group of) actor(s)'. For example, a piece of environmental law that lies down the emission or discharge limits for operations within industry sectors.

A major element of Rosenau's conception of authority is compliance. 'Authority is found in the readiness of those towards whom authority is directed to comply with the rules and policies promulgated by the authorities' (Rosenau, 2006, p. 175). Compliance is however not further defined, although it seems to refer to 'obeying or carrying out the wish or rules of others'.

Besides the lack of clearly defined concepts, other potential difficulties with the concepts rule systems and sphere of authority exist. First, concepts that define how changes in governance come about are absent. In fact, this is a broader problem in governance research. While changes are observed, concepts that allow for an analysis of how a change comes about are absent. This has as a result that even though spheres of authority and rule systems can be compared both within and across policy domains and over time, the difference between these spheres of authority and rule systems can not be explained.

Second, the use of Rosenau's conceptual framework would lead to a limited understanding of how the authority of the state (or other actors) can change. In this framework an actor either evokes compliance or not. This kind of authority is thus very much focused on what happens after the goals or objectives and the target group(s) of steering have already been defined. It is about making the other to abide the rules, to follow the procedures set out, to meet certain criteria, to reach the goal set, etc. However, I argue that having authority should not be delineated to having compliance generating capacities only, because having authority also means the ability to develop goals and measures that are regarded as legitimate by the target group to whom the goals and measures are directed. The phrase 'how has the authority of the state changed in governance' does not only refer to the outcome of political processes, i.e. the control and steering generated, but also to the process through which policies and steering are established. Therefore, the framework of Rosenau should be extended to include the development of legitimate goals and standards as a display of authority as well.

In other words, the focus on outcome in terms of compliance of the concept sphere of authority should be broadened in such a way that it includes the development of steering mechanisms as well. Related to that, while the sphere of authority does focus on multiple actors, levels and steering mechanisms, Rosenau's concepts do not incorporate multiple rules into the scope of analysis. Yet, focusing on political processes through which steering mechanisms are developed also requires an eye for rules of the game that allocate authority and that structure authority relations between actors.

In sum, the concepts of rule system and sphere of authority will be able to support the research done in this thesis because the two concepts emphasize steering, compliance and authority in governance. And to answer my research questions, I indeed have to assess the changes in this kind of steering capacity and authority of the state. At the same time, several difficulties have to be overcome, i.e. the lack of focus on the development of steering mechanisms and authority of actors in the political processes that shape this development and the lack of concepts to analyse how changes in governance and, in particular, authority come about.

2.3.2 Policy Arrangement Approach

Van Tatenhove and colleagues (2000b, p. 6) have developed the policy arrangement approach to study 'the interplay between 'conventional' and 'new' policy arrangements in different stages of the institutionalisation of environmental politics', but also to study the interaction between day-to-day environmental politics and structural social change, i.e. the changes in relationships between the state, the market and civil society. There are three concepts that are central in this approach: political modernization, policy arrangement and policy innovation.

The concept of political modernization refers to transformation processes in the political realm of society (Van Tatenhove, Arts & Leroy, 2000a). Traditionally, the political realm was defined by the rationalities of the state only, because the state, market and civil society used to have clear boundaries. However, one development that is associated with the emerging new spaces of politics is the blurring of the boundaries between these three subsystems. That is why the political realm in contemporary societies is not only characterized by the rationalities of the state, but increasingly also by the rationalities of the market and civil society (Van Tatenhove *et al.*, 2000a).

The policy arrangement concept 'refers to the temporary stabilisation of the organisation and substance of a policy domain at a specific level of policy making' (Arts, Van Tatenhove & Leroy, 2000, p. 54). The policy arrangement that characterizes a certain policy domain is on the one hand determined by a particular phase in the political modernization process, but on the other hand by interactions between actors involved in the day-to-day policy process. Just like Van der Zouwen (2006) notices, the phrase 'at a specific level of policy making' can best be left out from the definition of a policy arrangement if one focuses on multi-level political processes. The definition of a policy arrangement would then become 'the temporary stabilisation of the organisation and substance of a policy domain'.

Arts and colleagues (Arts *et al.*, 2000) also provide an operationalization of the concepts 'substance' and 'organization', which according to the definition are central aspects of policy arrangements. Substance, it is argued, refers to those concepts, ideas, views and buzzwords that give meaning to a policy domain. Substance is therefore operationalized in terms of the policy discourse in a certain policy domain. The organizational aspect of a policy arrangement knows more dimensions. The first is the actors and their coalitions that interact with each other within a policy domain. The second dimension is the power resources distributed among the actors and the power relations between them. The final dimension of the organization of a policy arrangement is those rules, procedures and norms that structure the behaviour of and the interaction between actors.

Even though this theoretical framework was originally presented as consisting of two main concepts, namely political modernization and policy arrangement (see Van Tatenhove *et al.*, 2000b), the authors introduced a third important concept in later articles, namely policy innovation (Arts, Leroy & Van Tatenhove, 2006; Arts & Van Tatenhove, 2004). Policy innovation is defined as the renewal of policy making in day-to-day interactions in arrangements and refers to doing things differently. The way that a difference in doing things can come about is either through the decisions of the actors involved in the policy arrangement or through a spill-over effect from another policy arrangement. Moreover, this change in how policy making is done, can be related to changes in both the substance or in the organization of the policy arrangement. In other words, such a change starts with a change in the discourse, actors involved, power resources and relations, or the rules of the game of a policy arrangement and then sets of a chain reaction in the other dimensions (Van Tatenhove *et al.*, 2000b).

The policy arrangement approach has been inspired by the structuration theory of Giddens (Leroy & Arts, 2006; Van Tatenhove *et al.*, 2000b), which deals with the duality of structure. This theory argues that the structure constitutes the act of an agent, but that at the same time the structure is reproduced or changed through the act of the agent. The act of the agent is therefore the place where structure and agency meet each other.

In the policy arrangement approach structure and agency meet each other in the policy arrangement, where actors act in the development and implementation of policies. Structure is represented by the concept of political modernization. Political modernization can therefore be used to analyse the structural features and changes in the socio-political environment in which the policy domain is embedded (structure). Political modernization in turn also structures the organization and substance of a policy domain to some extent (structuration). On the other hand, the actors in the policy arrangement also co-determine the organization and substance of their arrangement and of the policy domain (agency). These actors do that either by reproducing the structural elements of the policy arrangement or by initiating policy innovation through which the organization and substance of the policy arrangement can change.

The policy arrangement approach thus provides for a framework to analyse stability and change in policy practices in a policy domain and on a higher level within the socio-political realm of a society or across societies. Through the concept of policy innovation it also provides for the analysis of where changes in the policy arrangement or political modernization originate from. The opportunities that the policy arrangement approach provide for a study into changes in governance is confirmed by Van der Zouwen (2006, p. 227), who concludes – after having used this approach in her own PhD-research – that 'the policy arrangement approach enabled to study the emergence of governance and multi-level governance practices in more detail'.

Moreover, through the focus on organization and substance within the policy arrangement, the interaction of actors in the development and implementation of policies (steering mechanisms) can be analysed. In doing so, the approach also provides for actors from the state, market, or civil society to be involved in or at the centre of political processes. Moreover, contrary to the concepts of rule system and sphere of authority, this framework regards rules of the game, both established and non-established, as one of the dimensions that structure a policy arrangement. This focus on interaction between actors also allows for analysing changes in the involvement of actors, the discourse used by different actors, the power resources of actors and the power relations between actors, and the rules underlying the behaviour of actors in a policy arrangement.

Yet, the main drawback of the policy arrangement approach is its lack of focus on the steering mechanisms developed. The concepts political modernization, policy arrangement and policy innovation do not pay attention to the development and use of multiple steering mechanisms within a policy domain. The concepts do not deny the possibility of multiple steering mechanisms, but there is no focus on it either. In addition to that, the concepts do not provide for an analysis of whether the result of the political processes can be regarded as steering or not. Rather it is a silent assumption that where a policy arrangement exists, steering is being exercised. Yet, I would argue that steering is about developing legitimate policy and/or being able to generate compliance from the target group(s) and that these output and outcome elements should be part of a theoretical framework for governance.

Thus, it can be concluded that where the concepts rule system and sphere of authority lack a focus on the political processes in governance, the concepts of policy innovation, policy arrangement and political modernization lack a focus on the results (in terms of steering mechanisms and compliance) of political processes. In addition, where rule systems and sphere of authority lack concepts for studying changes in governance practices the concept policy innovation allows for an analysis of how changes come about. Yet, this difference in focus does make the two approaches complementary to each other. The next section will therefore merge the two approaches. In doing so, the concept sphere of authority will be taken as a starting point. The reason behind this choice is that the term policy has connotation to governing through state-based steering mechanisms and, as argued before, a conceptual framework for governance should avoid such connotations.

2.4 Spheres of authority in governance

The previous section showed that merging Rosenau's framework for global governance and Van Tatenhove and colleagues' policy arrangements approach should lead to a complete and appropriate theoretical framework for analysing governance in general and for this thesis in particular.

The main concepts in both approaches to (global) governance are rule systems, sphere of authority, political modernization, policy arrangement and policy innovation. Merging both approaches into one is not just a matter of taking all these concepts to form one conceptual framework. One of the difficulties is that some concepts might overlap in meaning with each others, e.g. sphere of authority and policy arrangement both have actors as one of the focal points. Moreover, some concepts are not always defined in a way that does justice to the concepts used and phenomena analysed in this research, e.g. authority has to do with more abilities than generating compliance.

In the first section, I will take the concept sphere of authority as the central concept in this conceptual framework. Using the policy arrangement approach, I will argue that the concept sphere of authority can mean much more than the meaning that Rosenau attaches to this concept. I will also argue that it is able to incorporate both the policy arrangement concept and the concept of rule system. I will then turn to discuss the main aspects of the concept sphere of authority in terms of its organization, substance and results. The concepts of policy innovation and political modernization will be discussed in paragraphs 5 and 6, respectively.

2.4.1 Redefining sphere of authority

While the concepts political modernization and policy innovation refer to structural change and daily changes respectively, the three concepts policy arrangement, rule system and sphere of authority are associated with *daily policy practices*. As discussed before, the policy arrangement approach refers to the organization and substance of policy practices, while sphere of authority and rule system refer to the results of these practices. It was also argued that combining these views on governance and these three concepts in particular, can provide us with a more comprehensive framework for studying change and stability in (the authority of the state in) governance. In this section, I will argue that it is exactly the sphere of authority concept that is able to integrate the focus on organization, substance and results of policy practices. However, before the concept sphere of authority can do that, it needs to be redefined.

The definition that Rosenau uses for the concept sphere of authority is 'the collectivities that have the authority to evoke compliance from those who are ruled'. As argued above, this conceptualization is a very narrow one. There are two reasons why I find the concept sphere of authority to be narrowly defined. First, Rosenau has a restricted understanding of authority, because authority is here confined to the ability to generate compliance only. Second, a sphere is conceived in a narrow way as well, because it only refers to a collectivity of actors. Taken together, a sphere of authority refers thus to nothing more than an arrangement of actors than can generate compliance.

However, I believe that the concept sphere of authority comprises much more than actors and compliance. As the concept of policy arrangement shows, an arrangement of actors is not only about actors, but also about the relations between the actors, how they behave within the collective, what they discuss with each other, and how each actor tries to determine both the composition of the collective and the result of processes within the collective. In other words, a collectivity is characterized by organizational and substantial elements. The collectivity that is part of a sphere of authority can thus be equated with a policy arrangement that is generating compliance, i.e. is displaying authority. The policy arrangement is the *sphere* in the sphere of authority. The definition of a sphere of authority would then become: the temporary stabilization of the organization and substance of a policy domain within which actors display authority.

If we turn to the conceptualization of the concept authority, we can learn from the Oxford Advanced Learner's Dictionary of Current English (1995) that authority is more than the ability to generate compliance; authority is having 'the power to give orders *and* make others obey [italics added]' and 'having the power to make decisions or take action'. In processes of governance, the power to make decisions or to take action is generally materialized in steering mechanisms such as policy, laws, programmes, projects, procedures, etc. The power to make others obey materializes in compliance with these steering mechanisms by the target group. Consequently, authority is here defined as the ability to take decisions through the development of steering mechanisms and/or the ability to generate compliance with these steering mechanisms. The definition of a sphere of authority would then become: *the temporary stabilization of the organization and substance of a policy domain within which actors take decisions through the development of steering mechanisms and/or have the ability to generate compliance with these steering mechanisms*. Figure 1 visualizes these elements of a sphere of authority and the relations between these elements.

This section started with three concepts, namely policy arrangement, sphere of authority and system of rule. The redefinition of the concept sphere of authority broadened the scope of the concept considerably. Its scope is even broadened to such an extent, that policy arrangements are now one element of a sphere of authority. The meaning of the concept policy arrangement has remained the same although it has now a stronger link with the output and outcome of policy practices. The concept of rule systems has not returned in the redefinition of the sphere of authority concept. Again, the scope of the concept sphere of authority was broadened to such

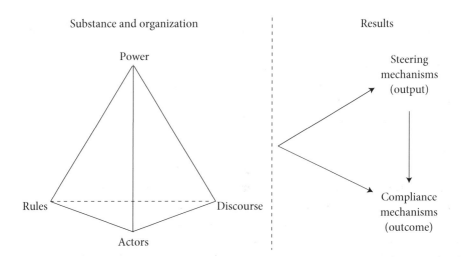

Figure 1. A sphere of authority and its dimensions.

extent that it has also incorporated the concept of rule system. After all, the development of steering mechanisms and authority that is generated through the steering mechanisms is now an element of a sphere of authority.

The broadening of the concept sphere of authority also acknowledges that it is not a choice between either the organization and substance that is important in analysing governance practices (like the policy arrangement emphasizes) or the results of governance (like Rosenau emphasizes), but that these three features are all aspects of a sphere of authority. The features organization and substance that come together in policy arrangements are now the sphere in the sphere of authority (see the left side of Figure 1). On the other hand, the results feature of a sphere of authority manifests itself in steering mechanisms on the one hand and compliance mechanisms on the other (see right sight of Figure 1). These three features will be further operationalized in the next section.

2.4.2 The organization of a sphere of authority

The first element of the organization of a sphere of authority is actors. In discussing this dimension of a policy arrangement, Arts and colleagues (2000) initially focused on how actors form coalitions. However, just like they have done in their most recent publication (Liefferink, 2006), I want to lay the emphasis on the actors themselves. There is no doubt that these actors will enter into coalitions during decision-making processes. However, the concern in debates on governance and thus also here is foremost with the plurality of the kind of actors that are part of governance practices.

One trend in governance is that not only governmental actors are part of the collectivity that develops goals and shape the implementation process of these goals, but that market and civil society increasingly become involved in these processes as well. In addition, since decision-making is increasingly a practice that takes place across multiple levels, the governmental, market and civil society actors involved in a sphere of authority can come from the subnational, national, regional, or international level. If we combine the trend of both different kinds of actors and actors coming from different levels, there is a whole array of actors that can become involved in policy practices. This array of actors is summarized in Table 1.

The second element of the organization of a sphere of authority is the dimension of power. This dimension refers to the distribution of power resources, the display of power and the power relations between the actors within the sphere of authority. This power is displayed during the policy practices in which these actors develop steering mechanisms and influence the implementation of steering mechanisms.

Power that actors have within a sphere of authority is based on the power resources they possess. These resources can be more formal in character, for example when they are based on finances or responsibilities based on formal rules and procedures. They have a more informal character when they are based on knowledge, verbal skills, etc. The distribution of resources is often unevenly spread among the actors. And the display of power is related to this interdependence that the uneven distribution of resources creates between actors (Klijn & Koppenjan, 1997). When one actor has the financial resources needed for the steering mechanism that is being developed within the sphere of authority, other actors that want to have the steering mechanism developed are dependent on this actor.

Still, there is a difference between having power resources and actually displaying power. Arts and Van Tatenhove (2004) argue that there are three types of power, relational power, dispositional power and structural power. Relational power refers to actors that are capable of achieving policy output through interactions. This type of power can either exists within a power struggle when actors want to achieve policy output against the will of others, or through a joint practice of a group of actors that wants to achieve a certain policy output together (Arts & Van Tatenhove, 2004). This type of power can be based on all kinds of power resources, such as information, rules and finances, which are used by actors within the interaction with other actors.

Table 1. An overview of possible actors involved in a sphere of authority.

	Governmental actors	Market actors	Civil society actors
Subnational	Municipality	Company based in one country, but producing in a particular community/province	NGO that operates locally, community, local group of citizens
National	Government, Ministries	Company based in and producing in one country	National NGO operating on the national level
Regional	Intergovernmental body with limited number of countries	Multinational, but production is located in a specific region	NGO based on a number of NGOs from countries within and operating in one region
International	UN, intergovernmental body with large number of countries	Multinational with world-wide production	International NGO with international membership and operating world-wide

The dispositional power type refers to power that actors have because they are positioned vis-à-vis other actors in a certain way. Actors are positioned vis-à-vis each other through the organizational structure within a sphere of authority and/or through the uneven distribution of power resources. This organization structure, the uneven distribution of power resources and subsequently the position of actors within this structure is mainly shaped by the rules that supply actors with their position and associated resources (Arts & Van Tatenhove, 2004).

Structural power also refers to power based on a structure, but this structure transcends the organizational level of the sphere of authority, because this type of power is based on macro-societal structures. This type of power comes from an asymmetric distribution of resources because of structural orders of signification, legitimization and domination within society (Arts & Van Tatenhove, 2004; Arts *et al.*, 2000). Moreover, because structural power is based on the macro-structures of society, this type of power is shaped by the political modernization in which the sphere of authority is embedded (see also Arts & Van Tatenhove, 2004).

The final element of the organization of a sphere of authority is the rules of the game. According to Arts and colleagues (2000) en Van der Zouwen (2006), the rules of the game establish the opportunities and barriers or possibilities and constraints for actors to act within the policy arrangement. In other words, rules regulate the behaviour of actors (Klijn & Koppenjan, 1997). These rules of the game can have a formal character, for example when there is a procedure or rule that is written down in a certain law or policy document. But the rules of the game can also be informal, for example certain norms or habits that shape behaviour. Moreover, certain rules can be part of a sphere of authority for a long time, while others are new and ad hoc and only have a momentary impact on the acts of actors in the sphere of authority.

Rules of the game are closely related to both the actor and the power dimension. Rules are linked with actors because rules influence the behaviour of actors, the position of actors in the sphere of authority and the interaction between actors. However, rules also provide power resources to actors when these rules allocate certain responsibilities or specific power resources to actors. Subsequently, rules shape the position of a certain actor and give this actor the ability to display dispositional power. In addition, both the rules and the resources provided by the rules can be used in the interaction between actors and are then part of the relational power of an actor.

2.4.3 The substance of a sphere of authority

The substance of a sphere of authority is shaped by the discourse dimension. Discourses are regarded as interpretative frameworks or dominant interpretative schemes, which give understanding and meaning to the policy domain (Arts *et al.*, 2000; Van der Zouwen, 2006). These interpretative schemes concern the framing of the problem and the possible solutions or criteria that solutions should meet. These interpretations consist of specific norms, concepts and terms; discourses are thus based on expectations, but are expressed through the jargon or language that is used by actors within the sphere of authority.

The discourse dimension is strongly related to the dimensions of actors, power and power relations and the rules of the game. Actors are the 'carriers' of the discourses, because they express and reproduce the dominant discourse through their language and during interaction with other actors. Discourses thus also shape the interaction between actors, because the discourse gives understanding and meaning to issues, the problem at hand and possible policy solutions. Discourses also define power relations between actors, because through a dominant interpretative framework and jargon certain solutions are pushed forward while other solutions are overlooked or neglected. Actors can therefore aim to change the existing discourse and through that change the framing of the problem and the solutions and the power relations that were based on the existing discourse. Finally, discourses are related to the rules of the game, because they provide rules with regard to the language used, the definition of the problem and the consideration of solutions.

2.4.4 The results of a sphere of authority

The results of a sphere of authority consist of two dimensions. The first dimension is steering mechanisms. Steering mechanisms are developed with the aim to change the behaviour of certain actors or to change certain processes in societies or organizations. Steering mechanisms are the instruments (mechanisms) through which this objective of change will be achieved. Steering mechanisms can be placed on a continuum ranging from having a voluntary to compulsory nature. For example, a norm like the precautionary principle can be a more voluntary steering mechanism, because it aims to increase the integration of environmental concerns in decision making by establishing the rule to give priority to environmental concerns even though there is uncertainty about the exact form and level of environmental damage. On the other extreme, an Environmental Protection Act that arranges the process through which environmental concerns are integrated in decision making and which industrial or societal activities and processes are subject to environmental regulation is a steering mechanisms as well and would be placed on the compulsory side of the continuum. In between these very hard mechanisms of formal rules and laws and soft mechanisms of steering such as norms, lies a whole array of mechanisms such as work programmes, voluntary agreements, standards, procedures, etc.

The other dimension that is part of the results of a sphere of authority is compliance mechanisms. Compliance refers to obeying or carrying out the wish or rules of others. Within the sphere of authority compliance means that the target group of the sphere of authority carries out the rules and objectives set by the steering mechanisms that are developed within the sphere of authority. The level of compliance can differ in each sphere of authority. Compliance can be anywhere between total defiance to automatic and immediate compliance; i.e. from total non-compliance to full compliance (Rosenau, 2003a). Generating compliance is done through compliance mechanisms which can range from formal (e.g. sanctions or inspections) to informal (e.g. norms or voluntary agreements) (Rosenau, 1992, 2002, 2003a). These mechanisms are the instruments through which compliance is achieved. For example, a formal compliance mechanism is inspections that the government does to control compliance with permits.

There is a direct link between the organization and substance of the sphere of authority and the dimension of compliance mechanisms (Figure 1), because the actors within a sphere of authority seek compliance from the targeted actor(s). However, compliance can also be linked to the steering mechanism within the sphere of authority, when compliance mechanisms are integrated in the steering mechanisms, for example through sanction for non-compliance, rewards for compliance, shared goals, work programmes. Steering mechanisms can contain compliance mechanisms to reinforce the instruments through which certain objectives are aimed to be achieved. In Figure 1, there is thus not only an arrow between the arrangement of actors and compliance mechanisms, but also between the steering and compliance mechanisms.

To conclude, the concept of sphere of authority allows for analysing the organization, substance and results of a policy domain at a specific moment in time. In this research, the concept will be used to research the authority of the state in two specific policy domains (shipping and offshore oil and gas production). This authority of the state has two aspects: the authority of the state vis-à-vis other actors during the development of steering mechanisms and the authority of the state vis-à-vis other actors in generating compliance during the implementation of steering mechanisms.

2.5 Changing authority in governance

According to the policy arrangements approach, change comes about through policy innovation. However, the concept of policy innovation analyses change in existing policy arrangements or spheres of authority, while at the beginning of this chapter it was argued that the main change in governance is considered to be a result of the emergence of new spheres of authority. When analysing changing authority in governance, one should thus look at both changes in existing spheres of authority and new spheres of authority. This matches the observation made in this chapter that when studying changes in governance both the still existing world of politics and the new world of politics should be taken into account. These two worlds of politics co-exist and interact with each other, which makes it vital for governance research to include both. The issue of policy innovation and the resulting renewal of existing spheres will be dealt with in Section 2.5.1, while the emergence of new spheres of authority is discussed in Section 2.5.2.

2.5.1 Renewal of an existing sphere of authority

Within the policy arrangement approach, policy innovation is defined as 'the renewal of policy making in day-to-day interactions in arrangements'. Yet, while the policy arrangement approach focuses on arrangements, I focus on sphere of authority. In this theoretical framework, policy innovation is therefore not only about the renewal of the substance and organization of a policy arrangement, but also about the renewal in the results of the sphere of authority, i.e. the authority that is displayed. The definition of policy innovation is thus rephrased to *the renewal of policy making and authority in day-to-day interaction in a sphere of authority*.

It should be noted, however, that the concept policy innovation carries several connotations that do not fit the new theoretical perspective on changing governance practices developed in this thesis. As also argued above, the term policy is most of the time associated with governmental steering mechanisms. In addition, innovation is often associated with improvements and modernization. Without doing away with the concept completely, the emphasis in this thesis will therefore be on changes in the dimensions of a sphere of authority that lead to a *renewed* sphere of authority.

In simple terms, the renewal of an existing sphere of authority results in doing things differently than before. Moreover, this process of renewal starts from and can affect all aspects and dimensions of a sphere of authority: it concerns the organizational aspect through changes in actors involved, the rules of the game and the power relations; it concerns substance when the discourse changes; and it concerns the results of a sphere of authority when the steering mechanisms or compliance mechanisms change. The influence of policy innovation on a sphere of authority is visualized in Figure 2.

The renewal of an existing sphere of authority comes about when a change in one dimension creates a chain reaction of change throughout other dimension resulting in a structurally change in the interaction between actors and/or the way in which steering mechanisms and compliance mechanisms are developed or implemented. Thus for example, if a rule of the game changes, but does not change the interaction between the actors or the results of the sphere of authority, the sphere of authority is not renewed. Likewise, the development of for example a new steering mechanism does not have to lead to a renewal of the sphere of authority. A renewal will only take place when the new steering mechanism leads to changes in the other dimensions and consequently to different results of the sphere of authority and/or changed interactions between actors in the sphere of authority.

The renewal of an existing sphere of authority through policy innovation is one source of change in governance. The process of renewal can be used to study in more detail how changes in the dimensions of a sphere of authority result in a change in authority of the state (or other actors). Moreover, the concept renewal helps to find those changes that are structural, i.e. the changes in governance and in the authority of the state that are institutionalized within the renewed sphere of authority. Temporary changes are therefore not part of the analysis.

However, there are two difficulties related to the analysis of the sources of and the institutionalization of changes in an existing sphere of authority that should be taken into account. First, because a renewal of a sphere of authority concerns a change in several dimensions and because it is expected that the changes in different dimensions will run parallel to each other, it might be difficult to demarcate from which dimension the renewal of the sphere of authority

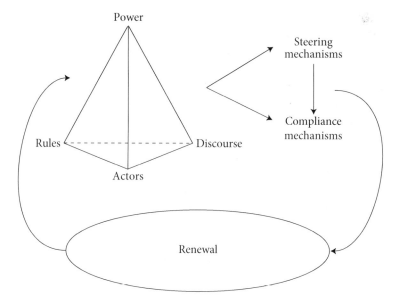

Figure 2. Renewal of a sphere of authority

originates. For example, a change in the rules of the game can change the role and responsibilities of actors simultaneously, resulting in a direct change of the power relations between the actors. Thus, a change in the rules of the game dimension can run parallel with a change in the power dimension, making it difficult to pinpoint to the origin of the renewal.

Subsequently, when a policy innovation in one dimension immediately sets off a change in another dimension, how can one determine in which dimension the change in the sphere of authority has started? This difficulty exists, because the dimensions of a sphere of authority are based on an analytical distinction, while they are – of course – very much interrelated in reality. This means that there are no clear criteria or indicators that tell where a policy innovation has started. The researcher therefore has to study its empirical material and has to justify its own interpretation of what the starting point and further process of the change in the sphere of authority has been through clear argumentation.

Another difficulty will be to determine when a sphere of authority is renewed and when the policy change is institutionalized in the sphere of authority (Van der Zouwen, 2006). The policy arrangement approach itself does not provide for criteria or indicators to assess when a change is institutionalized in a policy arrangement other then that the change should become 'temporarily stable'. But what is temporary; is it a week, a month, a year? This means that this assessment whether a policy innovation is regarded as institutionalized or not is subject to empirical research and to the interpretation of the researcher. This does not have to be problematic, as long as it is clear from the outset that the researcher is imposed with this task. Moreover, the researcher should be able to provide a clear argumentation for the choices he or she makes in this assessment.

2.5.2 New spheres of authority in governance

A second source of change in governance is the emergences of new spheres of authority. In fact, most literature on governance suggests that this is the most important source of change. While the analysis of changes in an existing sphere of authority presents some difficulties, the emergence of a new sphere of authority seems to be very straightforward. A new sphere of authority emerges when a new collectivity of actors develops steering and compliance mechanisms. Moreover, this collectivity of actors negotiates and shares (new) rules of the game, enters into power relations with each other, and shares and competes over the discourse that frames the problem and solutions that culminate in the steering and compliance dimensions. Finally, in order for it to be a sphere of authority, the collectivity of actors and the steering and compliance mechanisms it has developed should be relatively stable. The emergence of a new sphere of authority next to the existing one is visualized in Figure 3.

However, this in fact runs into similar problems as mentioned in the previous section. When is a collectivity of actors that displays authority institutionalized enough for it to be considered a sphere of authority? And when is the collectivity of actors and the steering and compliance mechanisms different enough from the existing sphere of authority to regard it as a separate sphere of authority? Again, there are no clear criteria for this puzzle. It is therefore up to the researcher to study and interpret the empirical material and to present it to others in a convincing way.

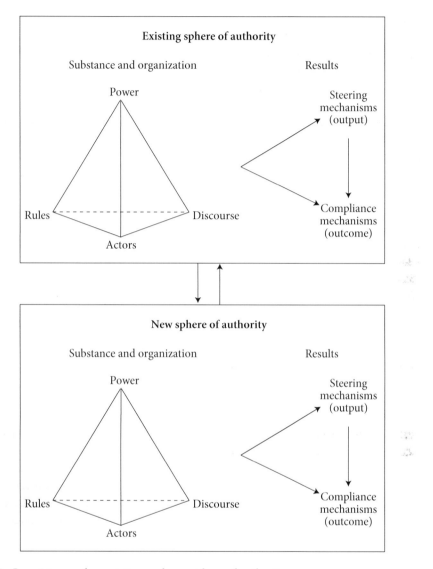

Figure 3. Co-existence of an existing and new sphere of authority.

2.6 Political modernization in governance

In a way, this chapter ends where it has started: with structural changes in how societies are governed. The final source of change in governance is structural change in the political landscape of a society or at the global level. The concept that the policy arrangement approach uses to refer to these structural changes is political modernization. This concept was originally defined as 'the transformation processes in the political realm of society' (Van Tatenhove *et al.*, 2000a). However, in this thesis, the policy arrangement approach is linked to the debate in governance and global governance.

In Section 2.1, it was argued that changes in governance manifest themselves in different properties, namely an increase in the diversity of actors that is involved in governance, the increasing spread of decision making along different governance levels, the increased diversity of rules that shape governance practices and an increased diversity in the steering mechanisms that are used within governance. That is why in this thesis, the process of political modernization will be delineated to these four shifts in governance.

Political modernization is a two-way transformation process; structural changes in the political realm have consequences for daily governance practices, but changes in daily governance practices also contribute to these structural transformations (Arts & Van Tatenhove, 2006). More specifically, the shifts in governance influence the organization, substance and results of a sphere of authority. At the same time, the shifts in governance towards multiple actors, levels, rules and steering mechanisms are either reproduced or changed through the governance practices that occur within a sphere of authority.

This thesis is able to contribute to the study of political modernization because of the long-term perspective that is taken in analysing the environmental governance of shipping and offshore oil and gas production. This thesis therefore allows for a reflection on the extent to which the two cases in this thesis have been affected by the four shifts in governance.

Whether the four shifts in governance as part of the broader process of political modernization within society is reinforced by the governance practices studied in this thesis is more difficult to assess. After all, a change in one policy domain does not necessarily mean a change in political modernization has occurred. The change in governance practices should be institutionalized in and across different spheres of authority in different policy domains, before it can be regarded as structural enough to be able to speak of a new phase in political modernization. In other words, for an analysis about changes in the structural features of a political realm of society, one needs to have evidence of structural changes in a whole set of policy domains, before one can conclude that there has been change in the political modernization of a society. Drawing conclusions about political modernization is therefore always difficult when it is based on a limited number of case-studies. The only domain where this thesis will be able to reflect upon is on political modernization within marine governance.

2.7 Conclusions: analysing the changing authority of the state in governance

One of the main arguments made in this chapter is the fact that changes in governance do not necessarily lead to an absolute shift from conventional governance practices to a new way of governing in new sites of politics. The shifts in governance do not make state-based or other conventional forms of governance obsolete. Rather, new governance practices will co-exist and interact with already existing governance practices. That also means that authority of actors in a policy domain is based on how authority is structured in the existing sphere(s) of authority and the new sphere(s) of authority.

In order to recognize the four shifts in governance that affect daily governance practices, but to also allow for the co-existence of existing and new governance practice, the concept sphere of authority was adopted. The definition of authority was broadened to allow compliance as well as the development of steering mechanisms as indicators of authority. To understand who has authority and why, the concept sphere of authority was also broadened to include the interaction between actors, which is based on the dominant discourses, the power resources and relations, and the rules of the game that guide the behaviour and interaction of the actors. Moreover, the renewal of existing spheres of authority and the development of new spheres of authority are considered the main ways in which changes in governance and in the authority of actors come about.

It is through the conceptual model developed in this chapter that the changes in the environmental governance of shipping and offshore oil and gas production will be analysed in this thesis. Moreover, based on the analysis of the changes in environmental governance, the changing authority of the state will be analysed as well. However, before this analysis can be done, I will reflect on some of the methodological choices I have made in this thesis in the next chapter.

Chapter 3.
Research methodology

3.1 Operationalizing the research questions

The marine environment and especially the maritime activities of shipping and offshore oil and gas production are interesting subjects for research into changing governance practices. The two activities have been subject to environmental governance for several decades already, the pollution generated by the two activities is by definition transboundary and because of the more intensive use of the sea, the number of stakeholders and their stakes in the environmental governance of the marine environment and these two activities are increasing. Moreover, the sovereignty of the state decreases the further one gets away from the mainland. The expectation is therefore that studying the environmental governance of these two activities can help in learning more about shifts in governance and the changing authority of the state. The objective of this thesis is therefore *to analyse how shifts in governance affect and change the authority of the state in the environmental governance of shipping and of offshore oil and gas production in order to generate new insights for the governance debate.* The following research questions have to be answered to achieve this objective:
- Which shifts in governance have taken place in the environmental governance of shipping and of offshore oil and gas production?
- How has the authority of the state in the environmental governance of shipping and of offshore oil and gas production changed under influence of shifts in governance?
- What are the insights gained from this research for the governance debate?

However, now that the conceptual framework for this thesis is developed, these research questions can be further operationalized. The previous chapter argues that in order to analyse to what extent shifts in governance have affected the environmental governance of shipping and of offshore oil and gas production, it is important to know the policy changes that the environmental governance of both sectors have gone through. After that, the effect of the policy changes on the sphere(s) of authority should be analysed. This effect consists of the emergence of a new sphere of authority and/or the renewal of the existing sphere of authority. By analysing the emerged new sphere of authority and/or the renewed existing sphere of authority, conclusions can be drawn about the extent in which the shifts in governance have affected the environmental governance of the two maritime sectors. The first research question is operationalized into three new research questions, reflecting the steps needed to analyse which shifts in governance have taken place in the environmental governance of shipping and of offshore oil and gas production:
- Which policy changes have taken place since the emergence of the first sphere of authority in the environmental governance of shipping and of offshore oil and gas production?
- What is the effect of these policy changes in terms of the renewal of existing and the emergence of new spheres of authority in the environmental governance of shipping and of offshore oil and gas production?

- To what extent has the renewal of existing and the emergence of new spheres of authority led to shifts in governance in the environmental governance of shipping and of offshore oil and gas production?

The second research question referred to the changing authority of the state in the environmental governance of shipping and of offshore oil and gas production. The conceptual framework makes a distinction between the authority of the state in developing steering mechanisms and the authority of the state in generating compliance. Furthermore, according to the conceptual framework, changes in the authority of the state are the result of policy changes. The research question is therefore operationalized into:

- How has the authority of the state in developing steering mechanisms and generating compliance changed as a result of the policy changes in the environmental governance of shipping and of offshore oil and gas production?

With regard to the last research question, which asks for the insights gained from this research for the governance debate, it should be noted that there are two ways in which this research aims to generate new insights. First, the development and application of new concepts to study changing governance practices should generate new insights. These concepts have been introduced in the previous chapter and will by applied in the coming chapters, which allows for a reflection on the use of the concepts at the end of this thesis. Second, based on the analysis of the environmental governance of shipping and offshore oil and gas production conclusions will be drawn about some issues that are discussed within governance literature, such as the nature of old and new governance arrangements, shifts taking place within governance and the changing authority of the state. This question is therefore further operationalized into:

- What are insights gained from the development and application of the theoretical concepts of this thesis and the results of the empirical analysis for the governance debate?

3.2 Case study and case selection

To answer the research questions just posed, in-depth analysis is required. In-depth insights are sought with regard to changing governance practices and changing authority of the state within these practices. The research strategy used in this research is therefore: case study. After all, as Marbry (2008, p. 214) states 'the raison d'etre of case study is deep understanding of particular instances of phenomena'. Moreover, to achieve deep understanding, empirical inquiry is done into 'a contemporary phenomenon in depth and within its real-life context, especially when the boundaries between phenomenon and context are not clearly evident' (Yin, 2009, p. 18). In this research, deep understanding of two particular instances is aimed for: the changing spheres of authority caused by policy changes and subsequent changes in the authority of the state. The phenomena in which these particular instances are studied are the environmental governance of shipping and of offshore oil and gas production.

According to Yin (1994, 2009), case studies can be exploratory, descriptive or explanatory in nature. For this research, five research questions have been formulated. The first, questioning the policy changes that have taken place in the two cases, is descriptive in nature. The answer to this question is a description of the historical development of the environmental governance of shipping and of offshore oil and gas production from inception to the present, so that the various policy changes can be identified. The second and third research questions ask what the effects of these policy changes are (the emergence of new spheres of authority, the renewal of existing spheres of authority and the occurrence of shifts in governance). These research questions are explanatory ones, because the answers require explanations about why the policy changes have caused certain effects. Similarly, the fourth question, which asks how the authority of the state has changed as a result of the policy changes, is also explanatory, i.e. the policy changes explain the changing authority of the state. Finally, the fifth question asks for new insights for the governance debate and in particular on the concepts used and issues studied in this thesis. This question is explorative in nature, because it entails exploring what the two cases in this research add to the existing debate on shifts in governance and the authority of the state.

In the introduction, it was already argued that from governing the marine environment, the maritime activities of shipping and of offshore oil and gas production were selected as cases for this thesis. One of the main reasons for choosing these cases is the long history they have with environmental governance. For a study into changing governance practices, it is interesting to have sectors that have been subject to environmental governance for some time already. The first environmental policies for shipping were developed in 1954 and for offshore oil and gas production in 1978. The long history will make it more likely that policy changes and changes in authority have occurred.

Another reason for taking these cases from the policy domain of the marine environment is that I expect this policy domain to serve as a breeding ground for governance initiatives. This expectation exits, first, because pollution in the marine environment is transboundary and therefore induces transnational cooperation. Second, because large parts of the oceans are a common and do not fall under the sovereignty of the state. Third, the scale and intensity of activities at sea are increasing, resulting in more stakes and stakeholders that have to be taken into account when governing the marine environment.

Finally, the fact that the cases of shipping and offshore oil and gas production are a relatively new policy domain for social sciences makes the cases appealing. Only shipping has been scrutinized in literature by authors from the field of international law. Other social sciences insights into how the environmental impact of shipping is regulated are lacking. About the environmental governance of offshore gas and oil production, there is hardly any social sciences literature at all.

Yet, there is one very important difference between shipping and offshore oil and gas production, which is worthy to note here. That is that ships are most of the time moving objects, while the platforms on which oil and gas is produced are static objects. The expectation is that this has induced differences in the sphere(s) of authority within the environmental governance of the two cases. This expectation is relevant because it helps in exploring the plethora of spheres of authority within the policy domain of the marine environment and it helps in gaining useful experience with the developed conceptual framework.

Another difference is the scale of the maritime activities. Shipping is a global activity, while offshore oil and gas production takes place regionally. Of course, offshore oil and gas production takes place in various parts of the world, but is concentrated in regions, while shipping takes place over long-distances and is spread around the world. Yet, in both cases, the focus in this thesis will be on the North Sea. There are several reasons for this focus on the North Sea. First, by taking a region as focal point, it will be easier to compare the shifts in governance and the changing authority of the state between the two cases. Second, the North Sea has several characteristics that make it an interesting region for studying changing governance practices in protecting the marine environment. For example, the North Sea is one of the busiest seas in the world (Halpern *et al.*, 2008), with many stakeholders and stakes competing for the use of the sea. Moreover, since the 1970s, several initiatives have evolved aiming to protect the marine environment of the North Sea (see for example Boehmer-Christiansen, 1984; Ducrotoy, Elliot & De Jonge, 2000; Skjaerseth, 2006). The expectation therefore is that the environmental governance of (maritime activities in) the North Sea have undergone several shifts in governance and subsequently changes in authority for the (North Sea) states.

3.3 Data collection

The main method of data collection used in this research is interviewing. A considerable number of interviews were done for each case: 21 for shipping and 30 for offshore oil and gas production (see Appendix A and B). The people that were interviewed were experts in the field of environmental governance for shipping or offshore oil and gas production. They derived this expertise from being involved in the environmental governance of one of the two maritime activities. Burnham and his colleagues (2008) even call this approach elite interviewing, because the balance of knowledge is in favour of the interviewee.

The selection of interviewees was also motivated by covering the entire field of the environmental governance of shipping and offshore oil and gas production, i.e. all spheres of authority involved in the two cases. Hence, a mixture of governmental officials, industry actors and environmental NGOs has been interviewed. The amount of interviews for each case was not decided a priori, but was a consequence of the point of saturation. When it was clear that information on all spheres of authority was gathered, no more interviews were done.

The result has been that a relatively high number of governmental officials have been interviewed for this research. In the offshore oil and gas case, governmental officials of those governmental agencies involved in regulating the offshore industry in the Netherlands, Norway and the UK were interviewed. These countries were selected because they are the three main countries having offshore oil and gas production in the North Sea. Moreover, I have interviewed governmental officials from the UK and the Netherlands for the shipping case as well, because both are countries with a long history in shipping. Besides that, governmental officials from the EU and IMO have been interviewed, since the environmental governance of shipping is concentrated on those levels.

Besides governmental officials, industry actors have been interviewed. Especially industry associations have been interviewed as they are the representatives of the industry. In the offshore oil and gas production case, individual companies were interviewed as well. This is related to the fact that in the environmental governance of offshore oil and gas production individual companies play a more specific role. Moreover, the number of companies is limited. Contrary to that, in the environmental governance of shipping, which largely takes place on the international level, individual companies are much less visible and there are a larger number of companies. These differences are also reflected in whom were interviewed in this research. Similarly, environmental

NGOs have also been approached for interviews, as representatives of civil society. Their involvement differs between the cases, something that is reflected in the number of environmental NGOs interviewed.

All in all, this approach also explains why a higher number of interviews were done for the case of offshore oil and gas production than for the shipping case. This former case exists of several spheres of authority that are involved in the environmental governance, while shipping has only one very dominant sphere of authority. As a consequence, covering the whole field for the offshore oil and gas production case automatically means doing more interviews than what was necessary for the shipping case.

Besides explaining who were interviewed, some remarks should be made about the way of working with regard to the interviews. First, all interviews were conducted face-to-face. Second, after the first interviews in 2005, a topic list was developed to guide the interviews. This topic list guarantees a level of consistency among the interviews. The interviews done for this research were, in other words, semi-structured interviews. Third, experience with the first couple of interviews learned that taping the interviews would allow for preserving the detailed information shared by the interviewees. The interviews made after March 2006 have therefore all been taped with permission from the interviewee. Fourth, all interviewees have received a copy of the transcripts made of the interview. If comments were given, they were incorporated in the transcripts. Finally, after the interviews were done, they were coded with the help of the program Atlas. The codes were developed based on the conceptual framework and the information provided by the interviews. This was especially useful for comparing the remarks of different interviewees on the same issue.

The second source of data is documents. During the research, efforts have been taken to get hold of documents from the most important meetings of the regulating bodies within the environmental governance of shipping and offshore oil and gas production. In particular, the summary records of the Marine Environmental Protection Committee of the International Maritime Organization and the Offshore Industry Committee (including its predecessors) of the OSPAR Convention since the mid-1980s have been successfully collected. These documents allow for a greater contextualization and balance (Burnham *et al.*, 2008). The documents therefore served as support for the information derived from the interviews. For example, based on the interviews, certain summary records were explored to provide supporting or additional empirical evidence.

The last and third method of data collection has been direct observation. Appendix C shows which meetings and seminars have been visited during this research. This method of data collection has especially been used to observe the governmental and industry relations, to learn what the relevant issues currently are in the environmental governance of the two maritime activities, to explore whether and how environmental NGOs were involved, and to meet people involved in the environmental governance of shipping and offshore oil and gas production.

3.4 Validity of the research

The issue of validity refers to the quality of the research done. For a case study, two types of validity are relevant, internal (also referred to as construct validity) and external validity (also referred to as generalization). Internal validity refers to the question whether you measure what you want to measure. One of the main ways to ensure internal validity is by data triangulation, i.e. collecting data from different persons or entities to be able to check the degree to which each source confirms, elaborates and disconfirms information from the other source (Marbry, 2008; Yin, 2009).

Data triangulation has been applied in this research. As explained in the previous section, the research underlying this thesis relies heavily on interviews. Yet, two other sources of data, i.e. documents and direct observation, have been used to confirm and complement the information derived by interviews. Only in the offshore oil and gas production case, this has been somewhat less successful, since direct observation has only been used scarcely. Moreover, in the shipping case, the three sources of data were also complemented by literature that reflected earlier research into the international environmental governance of shipping.

Furthermore, the internal validity of this research has been guaranteed by the fact that experts from the whole field of environmental governance of shipping and offshore oil and gas production were interviewed, such as governmental officials of different regulating bodies, industry actors and environmental NGOs.

Usually, concerns exist about the external validity or generalization of a case study. One of the reasons why this concern arises is because the generalization of research is often understood as statistical generalization from a sample to the population or universe. This type of generalization is not possible in case study research, because the results are context-dependent. For example, the results between the environmental governance of shipping and offshore oil and gas production will already differ, because the environmental governance of shipping takes place in a different context than the environmental governance of offshore oil and gas production. The results of this thesis can therefore not directly be generalized to other maritime cases, other seas or other environmental issues. Yet, case studies often aim to generalize to theoretical propositions rather than from the sample to a larger population. Case study therefore requires a different kind of generalization, so called analytical generalization (Yin, 1994). Of course, a prerequisite for generalization is that the research done is internally valid. As just argued, this prerequisite is reached.

The analytical generalization of this research is based on the conceptual framework for studying changing governance practices and the changing authority of actors presented in this thesis. This framework is a construction based on existing theoretical concepts, i.e. sphere of authority of Rosenau and the policy arrangement approach of Van Tatenhove and colleagues. Especially the latter has already been applied to a variety of cases within environmental governance (Arts & Leroy, 2006; Arts *et al.*, 2006; Van der Zouwen, 2006). This means that although the concepts are applied onto the environmental governance of shipping and offshore oil and gas production in this thesis, the conceptual framework can be applied on a broader range of cases, including other maritime activities (also outside of the North Sea region), other environmental issues, and maybe even other policy domains. In Chapter 7, after the application of the conceptual framework on the two maritime cases, I will reflect on the analytical generalization further.

Chapter 4
Changing authority in the environmental governance of shipping

4.1 The emerging global sphere of authority during the 1950s and 1960s

The environmental governance of shipping has always taken place on the international level. The first attempts to develop a global sphere of authority to govern the environmental impacts of shipping were already made in the 1920s and 1930s. In 1926, the United States of America (US) and the United Kingdom (UK), experiencing oil pollution on their beaches, proposed to adopt an international treaty to regulate intentional discharges of oil (Mitchell, 1994; Tan, 2006). They took this initiative to establish international regulations because otherwise they would have to adopt unilateral regulations which would create competitive disadvantages for their own shipping industry (Tan, 2006). However, the draft Convention was never formally adopted, also not in 1935 when the UK revived the proposal for an international treaty. A third attempt, undertaken by the UK in the 1950s, was more successful. In 1954, the International Convention for the Prevention of Pollution of the Sea by Oil (OILPOL Convention) was adopted, marking the first step in the development of a global sphere of authority in the global environmental governance of shipping.

4.1.1 The OILPOL Convention and the Intergovernmental Maritime Consultative Organization

A global sphere of authority that governs environmental impacts from shipping started to take shape when 32 countries adopted the International Convention for the Prevention of Pollution of the Sea by Oil in 1954. It was again the UK who took the effort to organize a conference with the aim to adopt an international treaty to regulate oil pollution from shipping. The UK and some other states experienced oil pollution on their beaches and especially the UK government was under pressure from domestic environmental NGOs to combat oil pollution from shipping. This time, the appreciation for such a steering mechanism was shared more widely because of the rise in the amount of oil transported by tankers and because of the increased experience of oil pollution (Mitchell, 1994).

The OILPOL Convention sets a limit for the discharge of oily wastes of 100 parts per million of oil in water within a distance of 50 miles from the coast. Oily wastes were generated through washing the oil tanks of tankers with water. The resulting oily wastes were discharged into the sea. The requirements of the Convention meant that tankers could still discharge oil as long as it was not within 50 miles from the coast. The only option that tankers had within 50 miles of the coast was to retain the oily wastes onboard and dispose them when visiting a port. Port reception facilities for oily waste were thus a necessary condition. Yet, since these facilities were costly, most states that did not experience any oil pollution themselves resisted mandatory requirements on this issue (Mitchell, 1994). In the end, the states agreed to the phrase that they have to 'ensure provision' (Mitchell, 1994; Tan, 2006). In 1958, the OILPOL Convention received the required number of ratification and entered into force.

The adoption of this first steering mechanism in the environmental governance of shipping is one important factor in the coming about of the global sphere of authority. The other important factor is the establishment of the Intergovernmental Maritime Consultative Organization (IMCO). The IMCO was established through the adoption of the United Nations Geneva Convention in 1948. The new specialized organization under United Nations' (UN) auspicious was to 'provide a machinery for cooperation among governments … and to encourage the general adoption of the highest practicable standards in matters concerning maritime safety and efficiency of navigation' (International Maritime Organization, 1998a).

The Geneva Convention entered into force in 1958; 10 years after its adoption. To enter into force, 21 states had to ratify the convention, of which 7 had to have a fleet of more than one million tons. And this did not happen till 1957, because some states with large shipping interests were suspicious towards the function and role of IMCO (M'Gonigle & Zacher, 1979; Tan, 2006). As Farthing (1993) explains, the states and their shipping industry were not prepared to have governmental involvement in the commercial, competitive, managerial and economic aspects of shipping. In those issues, the industry governed itself and this should remain so. The only area in which the shipping industry and governments accepted governmental involvement was in technical matters. Thus, it was accepted that IMCO was to provide an institutional infrastructure for the adoption of safety and anti-pollution standards, but it was not accepted that it would become a more political body; an implicit consequence of the fact that the IMCO would be a UN specialized agency and a potential consequence of its power to regulate the commercial aspects of shipping (Blanco-Bazán, 2004). The word Consultative in IMCO's name was meant to restrict the organization's role to purely advisory and technical matters and a tacit understanding emerged that IMCO's would not exercise its economic mandate (Farthing, 1993; Tan, 2006).

Moreover, while some states were afraid that the Convention would interfere with their national shipping industries and laws, other states felt that the Convention was written for the few countries that dominated shipping at that time (International Maritime Organization, 1998a). At that time, the states with the largest interests in providing international shipping services were among others: Greece, the Netherlands, Norway, Sweden, the United Kingdom and the United States, while the countries with the largest interests in international seaborne trade were among others: Argentina, Australia, Belgium, Canada, France and India (International Maritime Organization, 2008).

After IMCO was finally established, it became the secretariat that would enhance and facilitate the development of steering mechanisms and would provide a forum in which this could take place. As such the IMCO and the Geneva Convention provide this sphere of authority with formal rules of the game. These rules of the game define which actors participate in this sphere of authority and how new steering mechanisms will come about.

For example, IMCO consisted of an Assembly, a Council and a Maritime Safety Committee. The Council consisted of 16 members, of which 12 had to be with large interests in seaborne trade and shipping services. The Maritime Safety Committee, the body that took the most important decisions, consisted of 14 states of which 8 had to be the largest ship owning nations. The rules of the game also defined that the main function of IMCO would be to pass recommendations, convene conferences, draw up conventions and to facilitate consultations among member states (M'Gonigle & Zacher, 1979). Subsequently, IMCO could be characterized as a forum at which especially states with shipping interests met to discuss international steering mechanisms for shipping.

4.1.2 Freedom of the sea and flag states

As just explained, it was flag states that dominated the actor dimension of the global sphere of authority in the environmental governance of shipping in the 1950s and 1960s. Ships sail under the flag of a certain state, i.e. under the flag of the state where the ship is registered. What is more, this state was, at that time, the one and only actor that had jurisdiction, both in terms of setting standards and in enforcing those standards, over the ship. During the 1950s and 1960s, important flag states were Canada, Denmark, France, Germany, Greece, Italy, Japan, Liberia, India, the Netherlands, Norway, Panama, Spain, Sweden, United Kingdom, United States and Russia (Soviet Union) (DeSombre, 2006).

The primacy of the flag state is associated with the freedom of the sea principle. This principle was articulated by Hugo Grotius in the 17th century. Essentially, the principle means that the sea is for everybody and thus – except for a small stretch of three nautical miles along the coast – owned by nobody. Under the freedom of the sea principle, a ship is free to use the sea to go anywhere it wants to go and to carry anything it wants to carry. The only one that can set limits to this freedom is the flag state. Until 1982, the freedom of the sea principle was customary international law with regard to the sea and ships using the sea.

The freedom of the sea principle and the exclusive jurisdictions of flag states were thus important rules of the game in the global sphere of authority that emerged in the 1950s and 1960s. As such, these rules of the game are a very strong factor in explaining why states with shipping interests (i.e. flag states) are at the centre of the actor dimension of this sphere of authority.

Looking at the list of those states with the largest fleets registered in their country, it can be concluded that the shipping community was very much dominated by the developed world. However, three countries seem to be exceptions to that rule: Liberia, India and Panama. For two of these three countries, the main explanation of why they were among the largest flag states is because they have open registries for ships.

Open registries is a phenomenon that has been gaining popularity since World War II (DeSombre, 2006). Open registries do not have any nationality requirements with regard to the registration of a ship, i.e. ship owners, ship operators, crew, etc. do not have to be citizens of the country where the ship is registered. In other words, open registries do not require a genuine link between the ship and the state. Moreover, open registers attract foreign ship owners with low tax levels and relaxed standard and enforcement systems. These open registries are more commonly known as flags of convenience, because ships flying these flags usually have lower operations costs and/or less regulatory burdens.

The freedom of the sea principle, flag state primacy and flags of convenience do not only shape the actor and rules of the game dimensions. They are also associated with important discourses within this global sphere of authority. For example, important in the freedom of the sea principle is the word freedom. The word freedom brings with it certain connotations, for example that there are no limits. For the sea this means that the use of the sea is limitless and that regulatory limits over the sea or someone using the sea should be as little as possible. For shipping, if regulation is imposed onto ships at all, the only actor that is legitimate to do so is the flag state.

In addition, the term freedom also refers to the regulatory freedom in which shipping should be able to operate. The shipping industry and most flag states were of the opinion that most aspects of shipping should be regulated by the industry itself and not by governments. Only technical aspects such as the way a ship is constructed or operated are to be regulated – through IMCO – by governments. Thus another discursive element is the technical nature of both the scope of and the debate within IMCO. Subsequently the discourse within IMCO was largely confined to technological issues. Moreover, although IMCO had a pollution control side to its work because of the OILPOL Convention, most of its work in the 1950s and 1960s was on safety matters, which are very technical in nature as well. Examples are developing structural requirements for shipping, life saving equipment, load lines, etc.

Next to the technical nature of the discourse within IMCO, the term flags of convenience can be considered an element of the discourse dimension of this global sphere of authority as well. This term shows the attitude of the shipping community towards these open registries. What was especially disliked is the fact that these flags of convenience create a competitive advantage by having a convenient tax and regulatory system that leads to lower operational costs for ships (Barton, 1999).

A final element in the discourse dimension that emphasizes the extent to which flag states were central in this sphere of authority is the phrase that IMCO is a ship owners club. According to Blanco-Bazán (2004) this nick-naming was to a certain extent justified among others because of the way in which the flag states were institutionally embedded within IMCO. After all, flag states were brought to the fore through the way in which IMCO was organized. As mentioned above, certain rules of the game defined flag states as the main members of the Assembly, Council and Maritime Safety Committee.

To explain the reason why flag states operate in a global sphere of authority in regulating shipping, it is important to understand that shipping is a truly global economic activity. As several authors note (Barton, 1999; DeSombre, 2006; Marine Pollution Bulletin, 1975) a ship is built in one country, financed from another country, sold to a company in a third country and registered in a fourth. During a voyage it is chartered by someone from a fifth country, carrying cargo from a sixth to a seventh country, while belonging to a company in the eighth country. Thus, even though ships carry the nationality and fall under the jurisdiction of their flag states, in practice the issue of nationality is nonsense (Marine Pollution Bulletin, 1975). This explains the tension that exists between the global character of shipping practices and the primacy of the flag states in developing steering mechanisms for ships.

What is more, this activity takes for a large part place in an open access area, i.e. the sea. This is a direct result of rules of the game that define territorial waters to be limited to three nautical miles from the coast and the rest of the sea as a common. Subsequently, states are very interdependent when it comes to regulating shipping. Not only because environmental effects from shipping cross boundaries, but more importantly because those states experiencing pollution are not able to set those standards that are needed to protect their coast from pollution from shipping. Instead only a flag state is able to develop steering mechanisms to minimize the environmental effects from ships sailing under its flags.

Thus, the global nature of shipping and the fact that shipping takes place in an open access area makes it a footloose sector that is difficult to govern for a single state. But this is not all. One of the most important reasons why this is a footloose sector is that even though ships fall under the jurisdiction of a flag state, owners are able to register their ship in open registries (Barton,

1999). The result of the existence of these open registries is that if a single state chooses to adopt very stringent environmental standards (because it is experiencing pollution), it is likely that the state will lose part of its fleet to the open registries. These ship owners are not willing to have the burden of these regulations while other ships do not have this burden.

In other words, unilateral steering mechanisms only work counter productive, because they create a competitive disadvantage and an incentive for ships to flag to other states. In addition, unilateral steering mechanisms are not a viable solution to solve a problem like oil pollution from shipping, because those states experiencing pollution is dependent on the flag state to regulate the environmental impacts from shipping.

This is also the reason why the US and the UK have sought international cooperation to deal with the issue of oil pollution. Until the early 1960s, the US and the UK were the two biggest flag states having the largest fleets in the world. And they wanted to regulate oil pollution from shipping, because they were experiencing oil pollution themselves. On top of that, during the early 1950s, the UK government was under pressure from NGOs to do something about the oil pollution damaging beaches and birds. But unilateral action was not a viable solution for the UK, since it would create competitive disadvantages for the UK fleet and because it would not prevent other ships from polluting UK beaches. This explains why the sphere of authority that emerged in the 1950s and 1960s was a global one.

Thus the footloose nature of shipping is the main reason why the environmental governance of shipping has from the start been characterized by international steering mechanisms. However, the nature of shipping practices has also had its effect on the discourse dimension within the global sphere of authority. Both the footloose nature of shipping and the freedom of the sea principle make that the only viable way to deal with safety and pollution prevention issues is by developing international steering mechanisms and standards. The discourse is that national or regional approaches lead to unworkable situations and regulatory chaos for shipping. Moreover, national or regional approaches are not seen as legitimate, because they undermine the level playing field which is vital to the global activity that shipping is. Developing mechanisms is therefore confined to the IMCO only; IMCO has the exclusive mandate to bring flag states together and to set standards. As Farthing (1993, p. 161 (brackets added)) frames it 'the importance of international measure through im[c]o cannot be overstressed'.

4.1.3 Private actors

Even though flag states are important for the actor dimension of this sphere of authority, other actors, private actors, are also involved in the actor dimension of the global sphere of authority. First, environmental NGOs have been involved in the global environmental governance of shipping. The main role of environmental NGOs has been to pressure governments to take action against oil pollution from shipping (M'Gonigle & Zacher, 1979; Mitchell, 1994). As already mentioned, the UK led the call for international regulation for oil pollution and did so because it was under pressure to adopt unilateral regulations by the Advisory Committee on the Prevention of Oil Pollution (ACOPS), who established itself in 1952 especially for this purpose. Since the UK did not want to impose unilateral standards onto its fleet, the UK pressed for international action and organized the conference were the OILPOL Convention was adopted.

Second, industry actors have been part of this global sphere of authority as well. The interests of the ship owners have been represented by the International Chamber of Shipping (ICS) at IMCO since 1958. The ship owners are interested in developing steering mechanisms through IMCO, because they are afraid of having fragmented regulations on the national level instead (Interviews International Chamber of Shipping, 2007; International Maritime Organization, 2007).

Since the main target in this sphere of authority was oil tankers, the International Association of Tanker Owners (Intertanko) was involved as well. However, because the tanker association did not have observer status at IMCO, they allied with the ICS and represented tanker owners' interest through the ICS (Intertanko, 2001). It should be noted that the fact that industry associations were able to participate directly in the negotiations within this global sphere of authority is related to the rules of the game that allowed these actors to have observer states. Moreover, the reason why industry associations were involved in the negotiations is the input of the expertise of the industry (M'Gonigle & Zacher, 1979).

What is more, because the main issue targeted in this sphere of authority was oil pollution, the oil companies were also part of this sphere of authority. The oil companies also owned many of the tankers on which the initial regulations were focused. These industry actors were caught between the threat of unilateral regulations and their interests in having as little regulations as possible. Contrary to the environmental NGOs, the oil companies and tanker owners lobbied both at the domestic and international level.

4.1.4 Compliance

Yet, the footloose nature of shipping does not only limit the way in which the sector can be regulated, but also how compliance can be arranged. Since the flag state has full jurisdiction over its ships, the implementation of international standards and the generation of compliance with international standards is the responsibility of the flag state. However, ships are usually either somewhere out at sea or in a port and this port is most of the time not a port of the flag state. What is more, especially with open registries, a ship might not ever call at ports in the state where it is registered. Ships might be trading between Europe and Asia while being registered in Panama. In that case, a flag state never even has the ship within its own territory. There is thus a tension between the global nature of shipping and the primacy of flag states in ensuring compliance.

To overcome this tension, several compliance mechanisms have been developed. First, the OILPOL Convention granted states with ports the right to inspect ships which visit their ports, but under the condition that an inspection would not lead to a delay of the ship. In other words, the OILPOL Convention itself developed a compliance mechanism in which port states were granted with the authority to inspect ships in their ports.

The second compliance mechanism is the requirement that all operations involving oily wastes and discharges of this waste would be recorded in the 'oil record book' (Mitchell, 1994). Because of the footloose nature of shipping and the sometimes long-distance relationship between the flag state and the ship, the only place where enforcement is possible is while the ship is in a port. That is why it was agreed that states were allowed to inspect ships and their oil record books in their ports, as long as this did not cause any delay.

The third compliance mechanism was not directly focused at ensuring compliance from tankers, but from the flag states. This compliance mechanism is the only one that the IMCO secretariat has to enhance proper implementation, enforcement and compliance with international standards. States were required to report on reception facilities installed, the application of the treaty and actions taken on violations (Mitchell, 1994). However, according to both Tan (2006) and Mitchell (1994) reporting on the enforcement of OILPOL by individual states was virtually non-existent even though IMCO tried to push states to report through a survey in 1961 and a questionnaire in 1963. IMCO did not have any other means to compel flag states to report.

4.2 The authority of the state during the 1950s and 1960s

During the 1950s and 1960s, flag states were the only actor that had the formal authority to develop steering mechanisms. In contrast, flag states did not have the monopoly on authority in generating compliance. This section will analyse and explain the authority of these flag states during the development of steering mechanism in the 1950s and 1960s.

4.2.1 Authority in developing steering mechanisms

It should first be noted that the footloose nature of shipping in a way limits the authority of individual flag states in setting environmental standards for their ships and to protect their shores against ships' pollution. The footloose nature causes dynamics that prevent the development of national steering mechanisms and instead push states to agree on international standards instead. The only way in which states could retain their authority in developing steering mechanisms is by cooperating and setting standards which all flag states and their fleets have to apply. And flag states have indeed done so through the OILPOL Convention and by establishing IMCO.

In addition, this authority is shared with other flag states, because only internationally agreed standards by flag states are regarded as legitimate, by the flag states themselves as well as by the shipping industry. The question remains, however, what the influence is of these flag states and other actors on the steering mechanisms that are developed? Although the scope of this research prevents an in-depth analysis of the influence of actors on the content of steering mechanisms, some general observations can be made from analysing especially the actor and power dimensions.

First, environmental NGOs were part of the actor dimension of the global sphere of authority and they have been pressuring governments to take action against oil pollution from shipping (M'Gonigle & Zacher, 1979; Mitchell, 1994). Especially ACOPS has tried to influence the development of steering mechanisms by pressuring the UK. In 1962, the OILPOL Convention was amended. It was again ACOPS who drew attention to the issue of oil pollution. However, while environmental NGOs were largely responsible for putting the issue of oil pollution on the agenda, they had 'little direct input at the international level' (Mitchell, 1994, p. 107). This might have to do with the fact that during the 1950s and 1960s environmental NGOs were not a very common phenomenon yet and there were thus also no environmental NGOs with consultative status at IMCO.

The influence of environmental NGOs on the content of the steering mechanisms during the 1950s and 1960s has thus been indirect influence at best. Some NGOs have, after putting the issue on the agenda, pressured their governments to adopt regulations. This might have influenced the position of the governments during the international negotiations during the adoption and amendments of the OILPOL Convention.

The industry, however, has had more authority in the development of steering mechanisms during the 1950s and 1960s. Although it should be noted that according to M'Gonigle and Zacher (1979), until 1962, the oil and shipping industry had paid little attention to the creation of international regulations. Their authority since the 1960s is explained by two factors. First, they are able to influence steering mechanisms adopted since the interests of flag states are most of the time closely aligned with the interests of their shipping industry. For example, the flag states that opposed stringent environmental standards did so because they did not want to impose the costs onto their industries (M'Gonigle & Zacher, 1979). Unfortunately, the assessment whether this position of these flag states is related to active lobbying of the shipping industry or whether it was the position of the flag state from the outset requires the in-depth study of negotiations which goes beyond the scope of this research.

The second way, and probably a more important way, in which the shipping and oil industry has been able to exercise authority in developing steering mechanisms is by developing techniques and best practices. It was when the oil industry, and especially Shell, realized that the focus of governments were more and more shifting to combating oil pollution and that this would be done by imposing expensive requirements that they started to look for cheap alternatives (M'Gonigle & Zacher, 1979). These cheap alternatives in the form of techniques and operations have in many cases become the central part of the negotiations within the global sphere of authority and are often the basis for the compromise reached. In 1963, Shell discovered the 'Load on Top' procedure. The Load on Top procedure means that oily wastes are pumped into a separate tank where the oil will float to the surface. The water can then be discharged overboard, while the floating oil can be pumped back into the oil tank. Shell began to win the support of other oil companies and tanker owners and the use of Load on Top started to spread among the oil and tanker industry (M'Gonigle & Zacher, 1979). In doing so, they were able to make this procedure the focal point of the debates during the amendments of the OILPOL Convention in 1969.

Still, even though the environmental NGOs were important for putting the issue of oil pollution onto the agenda of several flag states and that the shipping industry provided important input into the negotiations for new environmental standards within the OILPOL Convention, the negotiations were for a large part shaped by states.

For example, in 1954, the debate revolved around whether to prohibit tankers to discharge oily waste when going to ports with reception facilities or whether to introduce a zoning system. While the UK together with a few larger flag states and states without a large shipping industry advocated the prohibition-option, the larger flag states, among other the US and some of the developing countries present advocated the zoning approach (M'Gonigle & Zacher, 1979). Behind the latter position lay motives such as lack of domestic concern over oil pollution, believing that oil would evaporate and dissolve, or the desire to protect the maritime interests against the burden of regulations (Mitchell, 1994). According to M'Gonigle and Zacher (1979), the differing perceptions of the cost and benefits of the two options primarily dictated the position of various countries. Overall, one can argue that flag states were concerned about the costs of the regulations, except when they were victims of oil pollution, which was the case with the UK, the Netherlands and Germany.

Similarly, in the negotiations for the 1962 amendments, the issue was whether to adopt a total ban for large new tankers to discharge oily waste as proposed by the UK or whether to expand the zoning approach. Again, some large flag states opposed the ban for new large tankers, i.e. Japan, the Netherlands, Norway and the US (M'Gonigle & Zacher, 1979; Mitchell, 1994). Still, the UK proposal had sufficient support to get adopted. The reason is that the problem of oil pollution began to take alarming properties; almost all European countries were now facing oil pollution on their beaches.

Finally, during the 1969 amendments, the issue was whether to allow the Load on Top method to replace the total ban for dischargers for new tankers or not. The negotiations on this issue took place against the background of the accident of the tanker Torrey Canyon on March 18 1967. The tanker Torrey Canyon hit a rock near the South-West coast of the UK and broke a few days later. The 120,000 tons of oil that the tanker was carrying contaminated miles of beach in the UK and France. 15,000 sea birds were killed. To make things worse, about 42 vessels sprayed 10,000 tons of detergents onto the floating oil in an attempt to emulsify and disperse it, but these substances were extremely toxic to many marine organisms and caused even more damage. The Torrey Canyon accident was the first tanker accident that caught public attention through broad media coverage. The accident raised public concern over oil pollution in many European countries.

Contrary to before, the UK government began working closely with its oil industry and therefore advocated the Load on Top procedure instead of discharge limits (Mitchell, 1994). The oil and shipping industry, Norway, the Netherlands and France also supported the Load on Top option (Mitchell, 1994). On the other hand, the US started to experience domestic environmental concern and wanted more stringent international discharge limits (Mitchell, 1994). The US was supported by other states, such as the Union of Soviet Socialist Republics, Japan and Germany (Tan, 2006).

The outcome was a compromise: load on top would be legitimized, but the 50 miles zone stayed in place and new discharge limits were set as well. These new discharge limits were no longer based on the content of oil in water, instead the rate of oil discharges would be limited to 60 litres per mile and the total discharge was limited to 1/15,000 of a tanker's cargo capacity. In addition, these limits applied to waters outside of the 50 miles zones from the coasts. Within these zones tankers were permitted to only discharge 'clean ballast'. Clean ballast relied on the visible determination that the discharged water does not contain any oil. Still, it would take to 1978 before the 1969 amendments would enter into force.

What can be concluded from the above description of the negotiations taking place over the years within this global sphere of authority is that the development of steering mechanism during the 1950s and 1960s is driven by flag states that are concerned about oil pollution damaging their beaches. They generally advocate stringent standards to combat oil pollution, but meet opposition from the states whose maritime interests prevail, from the tanker owners and from the oil companies. These actors are all much more reluctant to accept such stringent regulations, because they will have to carry to burdens and costs of implementing those regulations.

The power dimension of this sphere of authority is therefore shaped by the power struggle between the actors with maritime interests that want to minimize the additional burdens posed by environmental regulation and the actors with environmental concerns that want to combat the environmental impacts from shipping. The outcome of this power struggle for a large part defines the influence that coalitions, individual states and industry associations have had over the steering mechanism developed.

4.2.2 Authority in generating compliance

As explained in Chapter 2, the authority of the state in global governance is not limited to developing steering mechanisms only. Implementing, enforcing and ensuring compliance are the other side of the coin of governing. This section therefore explores the authority of the flag states in generating compliance in the global environmental governance of shipping in the 1950s and 1960s.

The footloose nature of shipping is limiting the way in which a flag state can generate compliance from its ships. But this is not the only reason why compliance with the OILPOL Convention was virtually non-existent. As both Mitchell (1994) and Tan (2006) explain the lack of compliance is also related to the nature of the standards set in the steering mechanism and the lack of compliance mechanisms to enforce and control such standards. Because, even though there were discharge limits for oily waste, there were no monitoring devices to assure compliance with this limit.

Furthermore, as just noted, rules of the game in the global sphere of authority have denoted flag states with the formal authority to enforce the OILPOL Convention. Yet, most flag states did not have the incentive to take appropriate efforts to control compliance (Tan, 2006), because it would cost them money while they were not the ones experiencing oil pollution.

This meant that compliance with this standard relied heavily on ship operators and their crew. However, they also lacked incentives to do so, because there was no enforcement by flag states. In addition, they lacked a financial incentive to recover oil, because independent tankers usually got paid for the oil loaded and not the amount delivered (Mitchell, 1994).

Two compliance mechanisms within the OILPOL Convention had the potential of generating compliance despite the reluctant flag states or tanker owners or crew. Namely, right of states to inspect ships which visit their ports and the oil record book. Yet, there are several difficulties that limit the effectiveness of these compliance mechanisms. The first difficulty is that port states had to gather the evidence of a violation which could then be presented to the flag state. However, it is extremely difficult to acquire sufficient evidence based on only an inspection in a port. After all, how can one be sure whether the ship has or has not discharged oil illegally within an area up to 50 nautical miles from the shore, when the only source of information is an oil record book?

Second, there is a certain interdependence between the flag state and the state of the port when it comes to making these inspections an effective part of the enforcement and compliance system in the global environmental governance of shipping. Because even though a port state could inspect ships, it still relied on the flag state for the prosecution of violations of the OILPOL Convention, because flag states had the exclusive jurisdiction to prosecute. The power to make port state inspections an effective compliance mechanism was therefore in hands of flag states. A port state was powerless when a flag state is not willing to cooperate. And again, flag states generally lacked the incentive to prosecute offending ships.

The problems with ensuring compliance by flag states, port states and the tankers themselves persisted throughout the 1960s despite the 1962 amendment after which new tankers became required to have oil content monitors and oily-water separators (because discharging oil was prohibited for them). Yet, these monitoring devices and oily-water separators did not exist yet (Mitchell, 1994; Tan, 2006). Still, these requirements spurred the oil companies to work on alternatives methods and operations to avoid the need for expensive (non-existent) monitoring devices and oily-water separators. Subsequently, during the negotiations for amendments to the OILPOL Convention in 1969, the requirement to have monitoring devices was traded for the requirements to adopt the Load on Top procedure. Thus in 1969, the requirements for monitoring devices were eliminated.

Still, improvements were made in the enforceability of the discharge limits because the general discharge limit became stricter. The discharges of oily water that leaves a visible trace would become prohibited within the zones. This made it easier for both ships and states to monitor compliance, because violations were easier to detect. On the other hand, the zones in which these visible traces of oil discharges were prohibited were still too large to actively patrol the whole area. What is more, since the Load on Top method is a procedure employed during the voyages, it is not easy to inspect that in ports. Thus even though generating compliance with the discharge limit of visible oil traces was easier to arrange, other standards within the OILPOL Convention were still as difficult to enforce as the 1954 standards.

It can thus be concluded that the effect of the stricter discharge limit on the total level of compliance and the enforceability of the standards was ambiguous. This is among others explained by the still existing rules of the game within the OILPOL Convention that did not allow port states to prosecute violators or make intrusive inspections which would make detection of violations more likely.

From the above, it can be concluded that during the 1950s and 1960s the enforcement system of the OILPOL Convention and the level of compliance were very weak. According to Mitchell (1994), this is largely due to the fact that the OILPOL Convention relied on a deterrent model of compliance, i.e. ensuring compliance was based on detecting non-compliance and prosecuting and sanctioning violators. However, as just explained, this does not work when those who have to comply (the ships) are constantly moving around at sea and when it is impossible to patrol the zones in which the discharge limits apply. The result was that compliance relied almost entirely on the integrity and self-incrimination of the ships crew and master. And that in a situation where not discharging oily waste and hand it in at a port reception facilities was made very difficult by the lack of adequate port reception facilities.

All in all, the enforcement of the 1954 OILPOL Convention was incredibly weak and flag states were not able to ensure compliance. In addition, port states were not able to fill the gap left by flag states. This means that even though formally flag states were the ones that have the most influence on ensuring compliance, in practice, states did not ensure compliance at all. Rather compliance was in hands of the ship owners and the crew. Yet, they did not have the incentive to comply or to develop their own compliance mechanisms. Important explaining factors for the weak authority of the state and the lack of compliance are the nature of the steering mechanism, the lack of incentives for both flag states and the industry to comply, the lack of effective compliance mechanisms and the power relations between the flag and port states.

4.3 The institutionalization of the port and coastal state during the 1970s and 1980s

As just argued, one important characteristic of the global environmental governance of shipping was the freedom of the sea principle and subsequently the primacy of the flag state. During the 1970s and 1980s, a policy change took place in which this flag state primacy was broken down. Through this policy change, port states and coastal states were institutionalized besides flag states as actors with their own interests, rights and obligations in the global environmental governance of shipping.

There were two developments in the 1950s and 1960s that laid the groundwork for the policy change that occurred between the early 1970s and the mid-1980s. The first development is that the formulation and adoption of global steering mechanisms focusing on reducing the impact of shipping on the marine environment was driven by (large) flag states with coastal concerns about oil pollution. The policy change that took place between the early 1970s and mid-1980s would ensure that these coastal concerns became institutionalized through the concept of coastal state.

Second, the OILPOL Convention created a distinct port state right by granting the opportunity to port states to inspect foreign ships. This was unprecedented until then. Yet, this port state right was limited to the OILPOL convention and the rights granted to port states were limited as well. The policy change discussed in this section ensured that the right of port states to inspect ships became more elaborated, became a widespread feature and marked the institutionalized role of port states within the global environmental governance of shipping.

4.3.1 The MARPOL Convention and port state inspections

The policy change started with the adoption of the International Convention for the Prevention of Pollution from Ships, commonly referred to as the MARPOL Convention, at a conference attended by 71 states in 1973. Just like the 1969 amendments of the OILPOL Convention, the adoption of the MARPOL Convention was directly related to the Torrey Canyon oil spill in 1967. Moreover, the adoption of this Convention was informed by the growing environmental awareness of the 1970s and the Stockholm Conference on the Human Environment in 1972.

The MARPOL Convention incorporated the OILPOL regulations in its Annex I, but focused on other kinds of marine pollution from shipping as well. The adopted Convention contained five annexes, each covering a specific source of marine pollution:
- Annex I Regulations for the Prevention of Pollution by Oil
- Annex II Regulations for the Control of Pollution by Noxious Liquid Substances in Bulk
- Annex III Prevention of Pollution by Harmful Substances Carried by Sea in Packaged Form
- Annex IV Prevention of Pollution by Sewage from Ships
- Annex V Prevention of Pollution by Garbage from Ships

Even though the adoption of this key steering mechanism is an important change in the global sphere of authority governing shipping, the adoption of the MARPOL Convention also brought about changes within the compliance dimension. The first aspect of this policy changes is that the MARPOL Convention introduced the International Oil Pollution Prevention certificate. The flag state had to inspect its ships regularly and issue this certificate as an evidence of compliance. The authority of the flag state in ensuring compliance was thus enhanced.

Second, despite the introduction of such a certificate, the states with coastal concerns and environmentalists still opposed these enforcement rules on the grounds that the record of flag state enforcement was unsatisfactory so far (Tan, 2006). During the negotiations of the MARPOL Convention, much more states were participating than in 1954. Among them were states with small fleets or states that were there to specifically defend their coastal interests, i.e. coastal states. Examples of coastal states are Australia, Canada, New Zealand, Ireland and Spain. Subsequently, a coalition of coastal states started to emerge (M'Gonigle & Zacher, 1979), which wanted more power for the coastal state to enforce international regulations.

This was, however, unacceptable for states with maritime interests and the Soviet bloc (Tan, 2006). Therefore the only rights the coastal state gained were to prevent a ship from sailing if it was presenting an unreasonable threat to the marine environment. In addition, it became mandatory for the coastal states to report violations to the flag state. In other words, flag state primacy in the compliance dimension was questioned by actors with environmental and coastal interests. Even though they were unsuccessful in bringing about major changes, they still managed to secure some compliance mechanisms for the coastal state.

But the most important new element of the MARPOL Convention in terms of enforcement and compliance was that the port state was granted with a more general right to inspect ships to verify discharge and equipment violations. If the ship did not comply with the requirements, these port states had to ensure that the ship would not sail until it can proceed to sea without being an unreasonable threat to the marine environment (Mitchell, 1994). But in doing so the port state was not allowed to challenge the International Oil Pollution Prevention certificate unless it had clear grounds for non-compliance. In that case the port state had to detain the ship until the ship could proof it posed no harm to the marine environment. Finally, the MARPOL Convention established the no more favourable treatment principle which means that all ships entering a port, even those of non-parties to the conventions, would be demanded to comply with the international standards (Blanco-Bazán, 2004).

According to Blanco-Bazán (2004) these provisions on port state control and the no more favourable treatment decisively altered Grotius' concept of freedom of the seas, because this meant an end to the exclusive jurisdiction of the flag state. The MARPOL Convention granted coastal states with some rights, expanded port state inspection rights and the scope of port state inspections to other pollution sources than oil.

4.3.2 The changing institutional structure of IMCO

The first change in the rules of the game that are the basis of the organization of IMCO was the adoption of the tacit acceptance procedure. This procedure permits the entry into force of amendments unless more than 1/3 of the contracting parties explicitly object. Before this tacit acceptance procedure, amendments of IMCO regulations had to be accepted by (usually) 2/3 of the contracting parties before they would enter into force (International Maritime Organization, 1998a). This explains for example the delay in ratification of the 1969 amendments of OILPOL 54, which did not enter into force till 1978. With the tacit acceptance procedure, many IMCO conventions, among others the MARPOL Convention, could be amended much more quickly and keep track with evolving technology and knowhow. The tacit acceptance procedure thus reduced the reliance on a majority of flag states to actively express their acceptance of amendments.

Second, in 1974, it was decided that the membership of the Maritime Safety Committee would be extended from 14 to all members. Before this change, the rule was that out of 14 members, 8 had to be the largest ship owning nations. With this institutional change, the bias towards representation of maritime interests was thus ended within the Maritime Safety Committee. This change was important in countering the criticism that IMCO was a ship-owner's club (International Maritime Organization, 1998a). This amendment to the Geneva Convention entered into force in 1978.

Third, the name of the Intergovernmental Maritime Consultative Organization was changed into the International Maritime Organization (IMO). These amendments to the Geneva Convention entered into force in 1982. The change in name from IMCO to IMO had several reasons. One reason is that by the mid-1970s it was clear that IMCO was more than a place where states met to consult, because IMCO had become a relatively well established treaty making organization (Blanco-Bazán, 2004).

Another reason is the development of the United Nations Law of the Sea between 1973 and 1982. One of the provisions of this Law of the Sea refers to accepted rules and standards adopted by competent international organizations. The fact that IMCO would be this competent international organization for shipping was not shared by everybody, because IMCO turned out to be little known and there were misconceptions with especially developing countries about the function of IMCO (Blanco-Bazán, 2004). This was related to the term intergovernmental in the Intergovernmental Maritime Consultative Organization and the dominant authority of ship owning states. To make IMCO a credible and competent international organization, the term intergovernmental had to be changed into international and ship owning states had to be put on equal foot with other states (Blanco-Bazán, 2004).

The fourth development in the rules of the game dimension of this global sphere of authority was that the adoption of the MARPOL Convention in 1973 led to a more explicit environmental mandate for IMCO. In 1975, a separate committee called the Marine Environment Protection Committee (MEPC) was established (International Maritime Organization, 1998a). Environmental matters were thus no longer dealt with in a subcommittee of the Maritime Safety Committee, but gained the recognition that they were a permanent part of the IMCO work program. This Committee is also open to all members of IMCO.

In addition, the growing environmental focus of IMCO was formally endorsed through a change in the aims of the organization as stated in the Geneva Convention in1977. Thus, the IMO was no longer 'to encourage the general adoption of the highest practicable standards in matters concerning maritime safety and efficiency of navigation', but would from 1984 onwards 'encourage the general adoption of the highest practicable standards in matters concerning maritime safety, efficiency of navigation *and prevention and control of marine pollution from ships*' (International Maritime Organization, 1998a).

All in all, with these formal institutional changes, the previous formal orientation of IMCO to facilitate cooperation of likeminded *developed maritime* states was abandoned (M'Gonigle & Zacher, 1979). Instead, IMO opened up and accepted states with coastal concerns and port states as equal members.

4.3.3 The United Nations Convention on the Law of the Sea

So far, the analysed changes in the compliance dimension of the global sphere of authority showed how the expansion of the rights of coastal and port states came about. The change in the organization of the IMCO showed how a first step was made in the institutionalization of the role of these states through balancing the representation of flag, port and coastal states in the institutional structure of IMCO. The final step in the institutionalization of these three types of states occurred, however, when the United Nations Convention on the Law of the Sea (UNCLOS) was adopted in April 1982.

With the adoption of UNCLOS, the freedom of the sea principle that had been dominant for centuries was replaced by a much more complicated juridical system of flag, port and coastal states. Negotiations for this Convention spanned a total of 9 years, i.e. 11 negotiation sessions were held between December 1973 and December 1982. The negotiations for this Convention were extensive, intensive and time-consuming, because of the global participation (166 members), the comprehensive scope of the work with the view to adopt a single convention on the law of the sea, the consultation with interest groups and the lack of a basic proposal for consideration (Jagota, 2000).

Besides arranging the width of the territorial sea, the Exclusive Economic Zone and the Continental Shelf, UNCLOS lies down the rights and obligations of flag, port and coastal states with respect to protecting the marine environment from pollution from ships. In that sense, the unfinished debates of jurisdictional claims during the development of the MARPOL Convention were brought over to and were resolved by UNCLOS (Tan, 2006). Moreover, UNCLOS provides for an important set of formal rules of the game with regard to flag, port and coastal states and how these states are involved in the global environmental governance of shipping.

Flag states still enjoy unlimited jurisdiction to prescribe standards and enforce these on their ships. The flag state is allowed to set more stringent regulations for their ships than the international requirements ask for, but has to take international requirements and standards as the minimum. It has to ensure the compliance of its ships with international standards. Moreover, the flag state must ensure that its ship carries the certificates that are required by international standards and that periodic inspections are made to verify whether these certificates accurately represent the condition of the ship. Nevertheless, due to UNCLOS the jurisdiction of the flag state is complemented with jurisdictions of the coastal and port states.

The fact that port states have some enforcement jurisdiction is not entirely new, because port state control and inspections were already part of the OILPOL and the MARPOL Convention. Nevertheless, UNCLOS has expanded the jurisdiction of port states considerably. As a general rule, port states enjoy full jurisdiction over a ship while it is in port and are thus allowed to set conditions and requirements for ships for entry into and use of the port. Moreover, before UNCLOS, a port state was allowed to inspect ships for violations of international standards and requirements in the internal or territorial waters of the state. However, UNCLOS expanded the right of inspections of port state to inspect ships for violations occurred on the high seas or in another state's coastal waters at the request of that state, the flag state or an injured state. Moreover, port states were granted the right to prosecute foreign ships that are in their ports for discharge violations of international regulations anywhere at sea. It is worthy to note that this prosecution right of port states was already discussed during the negotiations of the MARPOL Convention after a proposal of the US, Canada and Japan (Tan, 2006). At that time, however, it was rejected.

Still, even though the enforcement jurisdiction of port states is expanded, it remains optional instead of being mandatory. One of the limitations is that ports have to consider their viability and reputation; no port wants to acquire a reputation for being an overzealous enforcer of international standards (Tan, 2006). Because of the wide range of repercussions, it is unlikely that port states will pursue unilateral approaches (Molenaar, 2007).

The issue in which UNCLOS has also provided a breakthrough is the jurisdictions for coastal states. Coastal states' territorial sea was expanded from 3 to 12 nautical miles. Yet, the states with maritime and ship owning interests wanted to have their freedom of navigation assured within the territorial sea. The fact that coastal states gained control over larger parts of coastal waters jeopardized the freedom of the sea of ships. If coastal states could set requirements for their waters and thus also for ships sailing in their waters, ships would be faced with a plethora of different standards and regulations throughout their voyages. After all, they are likely to pass various territorial waters when sailing from one port to the other. It was unacceptable to both flag states and ship owners that the situation could emerge that a ship has to comply with completely different standards.

The conflict between coastal states and flag states on this point was especially focused on specific straits which allow ships to go from one sea or ocean to the other, e.g. the Strait of Gibraltar between the Atlantic Ocean and the Mediterranean Sea or the Strait of Dover between England and France connecting the North Sea and the English Channel. Coastal states were especially concerned about the passage of foreign warships so close to their shores.

The compromise that came out of the negotiations was the concept of transit passage, which gives ships (including warships) the right to navigation everywhere, also within territorial seas of other states, under the condition that they observe international regulations on safety and marine pollution. In all other matters, the straits are considered to be part of the territorial sea of the coastal state.

Within the territorial sea, the coastal state is sovereign. This means that coastal states have full jurisdiction to set standards and enforce them within their territorial sea. Nonetheless, because of the transit passage for ships, coastal states can only set standards that give effect to the international agreed regulations.

With regard to enforcement, the jurisdiction of coastal states is limited and subject to the provision that there are clear grounds for believing that a foreign vessel has violated conditions for access in a port or international regulations. When there are clear grounds for believing that a ship has violated certain regulations, the coastal state is allowed to ask the ship for information regarding its identity, flag, port of last and next call, etc. On top of that, the means for enforcing standards in the coastal areas is limited because coastal states generally do not have the manpower to watch and patrol these areas (Interview Ministry of Transport, Public Works and Water Management, 2005a).

The jurisdiction of coastal states is thus in particular restricted when it comes to ships transiting their territorial sea. Fortunately, when a ship is calling one of the ports of the coastal state, the jurisdiction of port states starts to apply and the state suddenly has more jurisdictions over that ship.

In describing the jurisdictions of the flag, port and coastal states, UNCLOS uses phrases like 'the safety, anti-pollution and seaworthiness standards established by the competent international organisation' or 'generally accepted international rules and standards'. It is generally believed that for shipping, the International Maritime Organization is the competent international organization to set the international standards to which UNCLOS refers (Blanco-Bazán, 2004; Frank, 2005; Tan, 2006).

Just like the development of UNCLOS took about a decade, the ratification cost a decade as well; UNCLOS did not formally enter into force till 1994. The reason for this long delay was not related to any controversy with regard to the jurisdictions set for flag, port and coastal states, but was related to the provisions it set for deep seabed mining (Jagota, 2000). That is why the system of flag, port and coastal states has been regarded as international law immediately after UNCLOS' adoption in 1982.

4.3.4 Memoranda of Understanding on Port State Control

As mentioned above, according to UNCLOS, port state inspections are a compliance mechanism that is optional to port states and not mandatory. IMO is widely regarded as the competent body to develop global standards for shipping within UNCLOS and within the shipping community in particular. For enforcement and compliance, however, this is not the case. To fill the still existing gap in compliance, several regional Memoranda of Understanding have emerged that have enhanced port state control in these regions and that have further institutionalized the authority that port states have in the global sphere of authority that governs the environmental impact of shipping.

The first Memoranda of Understanding on Port State Control was triggered by another major oil spill. In March 1978, the Amoco Cadiz ran aground near France. The Amoco Cadiz resulted in the worst oil spill ever; 223,000 tons of oil was spilled, covering more than a 100 beaches in France. The Amoco Cadiz spill prompted France to organize a conference with 14 European states, which took place in 1982. This conference focused on enforcement and ensuring compliance through inspections by the port states.

The outcome of the conference was the Paris Memorandum of Understanding (MoU) on Port State Control. This MoU is an administrative agreement between port states to coordinate and harmonize their port state inspections. The Paris MoU entered into force 6 months after its adoption in 1982 and has strong links with the enforcement regime established under the IMO regulations and UNCLOS, which both give the power to port states to inspect and detain ships.

The states agreed to inspect 25% of the ships entering their ports and to report the results of the inspection to a central computer processing facility. The requirements of the international conventions that have been agreed within IMO and the International Labour Organization provide the basis for these inspections. Initially, the inspections focused solely on monitoring equipment and certificate violations and not (yet) on operational discharge requirements (Mitchell, 1994).

The practice of port state control strengthened the enforcement and compliance of the OILPOL standards and the future MARPOL standards. Port states could make an important contribution to ensuring compliance from ships, because they can deprive flag states, owners or operators that make use of the primacy of flag state jurisdiction and act as a free rider or as a flag of convenience for the benefit of lower operation costs (Molenaar, 2007). The Paris MoU thus further institutionalized the rights and obligations of the port state in the environmental governance of shipping.

Yet, this reinforcement of enforcement was done at a regional level and not internationally within IMO. This caused unrest within IMO, because it was seen as 'discriminatory enforcement' of the international conventions (Blanco-Bazán, 2004, p. 281). As Blanco-Bazán (2004) further explains, it caused concern about imposing additional conditions for the entry into a port, but it also raised a debate on whether control procedures and enforcement were an exclusive IMO matter as well. The Secretary-General ended this discussion by emphasizing that IMO has a mandate to develop international standards and that port states are allowed to establish guidelines about implementation and enforcement of those standards. In the course of the 1990s, the Paris MoU would be supported by IMO by a resolution that called for similar mous to be adopted by other regions. And indeed other regions followed suit and adopted mous on Port State Control.

4.4 The renewed global sphere of authority during the 1980s and early 1990s

While the institutionalization of port and coastal states started in the rules of the game dimension and has affected the compliance dimension, this section will show that other dimensions of the global sphere of authority have been affected as well.

4.4.1 Ratification of the MARPOL Convention

The MARPOL Convention that was adopted in 1973 superseded the OILPOL Convention, but has a much broader scope as it includes pollution from the transportation of noxious and harmful substances, sewage and garbage as well. To become a party to the MARPOL Convention, a state only had to ratify Annex I and II; the other Annexes are optional. To enter into force, the MARPOL Convention required ratification by at least15 states with a combined merchant fleet of at least 50% of world tonnage. In the following years, however, only three countries ratified: Jordan, Kenya and Tunisia representing less than 1% of world tonnage (International Maritime Organization, unknown). This slow progress in the ratification of the MARPOL Convention was related to resistance against some requirements in Annex I and the more contentious and expensive Annex II (Mitchell, 1994). Some of the technology that Annex I called for did not exist yet, e.g. monitoring equipment that is associated with the Load on Top method, and Annex II covered many types of chemicals making it extremely difficult to implement (Tan, 2006). In addition, contrary to the OILPOL Convention, port reception facility were required by the MARPOL Convention, an expensive undertaking for many states (M'Gonigle & Zacher, 1979).

But in the winter of 1976 and 1977, in a period of a mere two months, about fifteen accidents with ships occurred in or near waters of the United States. In May 1977, the US took the lead in asking the IMO Council to adopt further regulations on tanker safety. The Council subsequently agreed, after a threat of the US to take unilateral action, to organize a conference on Tanker Safety and Pollution Prevention in 1978 (Tan, 2006). Delegates of 61 states attended this conference together with 17 international organizations (International Maritime Organization, 1998b).

The outcome of the conference was the adoption of a Protocol to the MARPOL Convention. To speed up ratification, the protocol established the provision that Parties to the MARPOL Convention would not be bound by Annex II for a period of three years from the date of entry into force of the protocol. The MARPOL Convention, i.e. annex I, entered into force in October 1983, a year after the sufficient number of states had ratified the convention.

4.4.2 Flag, port and coastal states

As we saw in the previous section, the renewed global sphere of authority shows several new rules of the game, most notably those within the IMO and the new rules of the games stemming from UNCLOS. These changed rules of the game have affected the actor dimension, because even though there is still a prominent place for states, a distinction can be made between whether a state is a flag, port or coastal state.

Still, it should be noted that this distinction is not always easy to make. There are several states, among others the US, the Netherlands and the UK, which are an important flag state, have a number of big ports and have a long coastline as well. However, these three types of states and their specific interests do not necessary have to be mutually exclusive. A state can have good environmental credentials and still be an important flag state. However, in these cases, a state might expect more from the environmental performance of its fleet than other flag states do (Interview Ministry of Transport, Public Works and Water Management, 2005a). Moreover, this distinction becomes easier to make when one looks at specific policy practices and the way in which certain initiatives, activities and interests of these states are related to either having a fleet, having ports or a long coast and how this has shaped the process or result of that particular policy practice.

In addition, coalitions have formed around these different types of states. As M'Gonigle and Zacher (1979) explain, within IMO this happened already in 1973 during the negotiations of the MARPOL Convention. These coalitions were informed by the upcoming Conference on the Law of the Sea. The different coalitions were coastal states, maritime states, developing states and the Soviet bloc.

Since the 1950s and 1960s, the states that were involved in these coalitions have, however, changed. In the 1950s and 1960s, it was countries like France, Italy, Japan, the Netherlands, Norway, Panama, Russia, the UK and the US that had the largest fleets. In the mid 1980s however, several new flag states had evolved, such as the Bahamas, China, Cyprus, Greece and Liberia. From the traditional flag states, especially the UK, the US and the Netherlands had already lost a large part of their fleets. For these states and some others (i.e. France, Italy and Russia) the role as port and coastal state has become much more important than their direct maritime interests (Interview Ministry of Transport, Public Works and Water Management, 2005a). By the end of the 1990s, the most important flag states were the Bahamas, China, Cyprus, Greece, Japan, Liberia, Malta, Norway, Panama and Singapore (DeSombre, 2006)

The reason why countries like Panama, Liberia and the Bahamas were able to acquire such a large fleet is not because they traded so many goods with other countries, but because they had created open registries as of the 1920s. Over the years, they attracted a lot of ships with their favourable conditions for registry, low taxation and more relaxed crew standards. In the early 1990s, 50% of the world fleet was registered under a Flag of Convenience.

The so-called North Sea Ministerial Conferences (NSMCs) are related to the establishment of coastal states as actors within the environmental governance of shipping. Just after the MARPOL Convention entered into force, the first NSMC was held in Bremen in 1984 (see North Sea

Ministerial Conference, 1984). Germany decided to convene a Ministerial conference with Ministers from all North Sea states and the European Commission, because it was concerned about the pollution of the marine environment and not satisfied with existing international regulations. The conference was meant to be a one-time event, but became a regular feature in the global environmental governance of shipping and the North Sea instead. Between 1984 and 2006, six Conferences and two intermediate meetings have been held.

The aim of these NSMCs was not to develop another international agreement but to develop a political declaration that could provide a political impetus to the already existing international agreements and institutions that focused on the protection of the marine environment of the North Sea. During the Conferences various issues were discussed, e.g. radioactive discharges, land-based sources of pollution, pollution from ships and monitoring.

The topics discussed with regard to the latter largely followed the topics under debate within IMO, but also demonstrated the agenda of the North Sea states. For example, the establishment of the North Sea as a special area under several of the MARPOL Annexes was a recurrent topic. The North Sea states would agree to pursue this at IMO together. The introduction of new topics or further amendments of MARPOL were also debated.

This means that the NSMCs were an important instrument in building coalitions between the North Sea states. The NSMCs served as a forum to discuss coastal matters and concerns that the North Sea states shared (Interview International Chamber of Shipping, 2007). Obviously, the consequence was that the North Sea states became a coalition of coastal states on several topics within the IMO.

For example, putting the quality of fuel and air emissions on the international agenda was one aim discussed during these NSMCs and subsequently pursued at IMO. The majority of ships use heavy fuel oil as a fuel for the ships. Heavy fuel oil is the residue of crude oil after the distillation process through which distillate fuel are produced. Heavy fuel oil therefore contains relatively high levels of sulphur. The burning of heave fuel oil causes considerable SO_2 and NOx emissions. Localized acid rain problems, due to these SO_2 emissions, were already raised in Europe and North America in the 1980s (Tan, 2006). The issue of air pollution and the quality of heavy fuel oils was brought up by Norway during the Second North Sea Ministerial Conference in 1987 (Tan, 2006). The debate and political declaration of the NSMC led to the issue also being raised at IMO (Tan, 2006). During MEPC 26 in September 1988, it was again Norway that proposed to include the quality of heavy fuel oils and air pollution in the future work program of the MEPC (Marine Environment Protection Committee, 1988).

Next to the change in the actor dimension that is related to the institutionalization of port and coastal states, environmental NGOs have started to participate with in the IMO as well. The four main environmental NGOs participating in the global sphere of authority were Friends of the Earth International (FOEI), the International Union for the Conservation of Nature and Natural Resources, Greenpeace and the Advisory Committee on Pollution of the Sea. According to a survey held among MEPC government delegates, the most important activities of these environmental NGOs were: submitting information documents to the meeting, their presence at meetings, their active participation in especially working and drafting groups and their submission of proposals to the MEPC (Peet, 1994).

4.4.3 The level playing field discourse

The new rules of the game, especially the fact that port and coastal states have certain rights to set standards for ships has affected the discourse dimension as well. First, the discourse that IMCO is a ship owners club has become less true, because the organization has opened up to a wider range of influences (Interview International Chamber of Shipping, 2007; M'Gonigle & Zacher, 1979). This, however, has not always refrained non-maritime actors from still nicknaming the organization as a ship owners club.

Other effects on the discourse dimension are, however, different in nature than one would expect. It was observed that the principle of freedom of the sea was associated with discourses on avoiding national or regional standards that create competitive (dis)advantages. Global standards are therefore the norm in the global environmental governance of shipping. Since UNCLOS and especially the concept transit passage replaced the freedom of the sea principle, one could expect a change in discourse as well. Remarkable, the discourses itself did not change. However, the importance of these discourses for the actors with maritime interests has changed. They have become more important precisely because they are no longer supported by the freedom of the sea principle.

Under the freedom of the sea principle, the discourse on maintaining the level playing field within the shipping industry first only applied to flag states not imposing unilateral regulations causing competitive disadvantages for their own fleet. However, because of new coastal and port states' rights and obligations, this level playing field now also applies to port states that should not impose regulations that would create competitive disadvantages for those ships visiting the ports of the port state. In other words, the potential for breaking down of the level playing field grew, because more states could cause distortions by imposing national regulations.

The consequence is that the call for regulating shipping on the global level has become more prominent as well, especially for flag states and other actors with maritime interests. That is why the discourse on maintaining the level playing field and the discourse on global standards have become even stronger after the adoption of UNCLOS. Indeed as M'Gonigle and Zacher (1979, p. 261) put it: 'for maritime interests, 'uniformity' has been the battle cry in a struggle to retain free access to all the world's oceans and to prevent the imposition of uneven costs'

This is most evidently exemplified by the way in which unilateral action of the US in imposing double hulls on oil tankers has led to the adoption of this requirement on the international level. The US introduced this requirement after the Exxon Valdez ran aground near Alaska and spilled about 36,000 tons of oil in March 1989. Shortly after the Exxon Valdez accident, the US adopted its Oil Pollution Act of 1990. This Act required all tankers calling at a US port to have double hulls by 2015. The only reason why the US was able to set these requirements and have a broader impact than just its own fleet, was because the US was now allowed to set conditions for entering a port as a port state.

This unilateral action was induced by domestic pressures. Environmentalists had been demanding double hulls for the safe transport of oil from Alaska for years, but this demand was never given into, among others because of the rejection of double hull requirements within the IMO in 1973 and 1978 (Tan, 2006). This laid the groundwork for the 'swell of public outrage against the oil and shipping industries' after the Exxon Valdez accident (Tan, 2006, p. 141). The public demand for double hull tankers in the US became even more determined when in February 1990 another tanker caused pollution in the waters of California.

By settling the double hull issue domestically, the US basically forced the IMO to start the discussion on the double hull immediately. Otherwise, the undesirable situation in which ships trading with the US are disadvantaged competitively would persist. The Netherlands and the Scandinavian countries supported the US proposal and were determined to adopt the requirement for both new and existing tankers (Tan, 2006). One of the proposals was to draw up a schedule for the upgrading of existing single hull tankers to double hull tankers.

The industry and developing countries strongly resisted the requirement for double hulls for existing tankers and had several reasons to do so (Tan, 2006). First, the industry argued that double hulls were not the answer to prevent accidents such as the Exxon Valdez. Double hulls would only be effective in accidents during low speed and grounding of a ship. During high-energy collisions both the outside and inner hull would be breached resulting in a spill of cargo. Second, the capacity of the global shipyard and ship repair would be insufficient to accommodate the demand for upgrading existing tankers to double hull tankers. Third, the capacity of the oil tanker fleet relied for 70% on single hull tankers built between 1973 and 1977. Any schedule for phasing out single hulls based on the age of a tanker would jeopardize the capacity in the fleet, while the global demand for oil was still growing.

Because of these strong industry concerns, acceptance grew that any schedule for the phase out of single hull tankers should take into account the transport demands for oil and the shipyard capacity to provide replacement tonnage (Tan, 2006). In addition, other designs equivalent to the double hull design in terms of safety were considered, especially after Japan reported the outcome on a study comparing three methods of design (Peet, 2008). The flexibility in terms of tanker design was advocated by countries with shipbuilding interests, such as Japan, Norway and France (Tan, 2006).

In 1992, Annex I was amended by adding the requirement of retrofitting existing tankers with a double hull within 30 years after their date of delivery and the requirements of a double hull or an alternative design that ensured the same level of protection against pollution (in the event of a collision or stranding) for new tankers. These amendments entered into force in July 1993.

Ironically, the US did not become a party to these amendments, because of the permission to use alternative designs (Tan, 2006). This was of course very disappointing to the other parties, because the US' demand for double hulls induced the amendments adopted in 1992. Still, the double hull requirement also applied to the US, albeit not through Annex I but through US' Oil Pollution Act of 1990. What is more, since the adoption of these regulations, there have been no orders for the construction of a ship using an alternative design (Peet, 2008).

Another unchanged discourse element is the fact that IMO is the only appropriate forum to develop steering mechanisms (Interviews Department for Transport, 2007; Friends of the Earth International, 2007; International Chamber of Shipping, 2007). This discourse is very much related to the call for global standards as the only viable way to regulate shipping. Furthermore, this discourse is enhanced because of UNCLOS and its implicit rule that IMO is the competent international authority to develop regulations for shipping.

Finally, the technical nature of adopted international standards and the technical nature of the debate in IMO still continue to characterize the work of IMO. As M'Gonigle and Zacher (1979) explain, the new participants and influences have been accused by the traditional maritime actors of politicizing the debate by introducing extraneous issues. However, at the same time, these maritime actors are able to have their gradualist approach to regulations reflected in the draft conventions. Moreover, 'an aura of sanctity surrounding the draft convention' prevents the adoption of substantial changes during the final conferences. Thus, even though the negotiations might have seen more politicized aspects, the overall nature of IMO's work is still technical in nature.

The emphasis on the technical nature of IMO's work is, just like it was in the 1950s and 1960s, related to the idea that shipping should be able to operate without being hampered by regulations and politics as much as possible. This part of the freedom of the sea principle thus still continued to exist. The regulatory freedom of shipping should be uphold as much as possible and only technical standards are sometimes necessary to impose to deal with safety or anti-pollution issues.

4.4.4 The threat of unilateral standards

The negotiations on the single/double hull issue also shows that with regard to the power dimension, the power dispute remains between environmental interests on the one hand and maritime interests on the other hand.

One of the strong power mechanisms used in negotiations within IMO are threats of unilateral action. These threats are used by states that want certain environmental standards to be adopted to protect the marine environment and their own coastline. Such a threat was not only used in the double hull issue, but also during the MARPOL negotiations in 1973 and 1978. Then the US threatened with unilateral action to break the resistance of maritime actors against the adoption of new global standards that would require changes in the construction of a ship and would bring about a reduction in oily wastes.

The difference is that back then, the US made this threat as a flag state, while in the double hull issue it made it as a port state. And the threats of port states to set certain requirements for entering ports are more powerful, because such entrance requirements have a more widespread effect on the whole shipping industry. Moreover, if a flag state imposes extra requirements, a ship owner can choose to register its ship in another state, where these extra requirements do not apply. Avoiding certain ports, especially those of the economic powerful states like the US, is much more difficult.

Still, the power of maritime actors, flag states and the industry remains strong as well. First, because some of the main flag states have close alliances with their shipping industry and thus represent the interests of that industry (Interview European Commission, 2007b). Second, because these states are backed up by several industry associations that have observer status. During these years several environmental NGOs became observer within IMO. Yet, the industry associations are much larger in number and thus remain more influential (Peet, 1994). Third, the argument that global standards are the best way to regulate shipping is a powerful one that in general every member of IMO agrees with. Thus port and flag states that want to protect their environment will not very easily go down the route of imposing unilateral environmental standards. Subsequently, if an issue is then dealt with within IMO, the maritime actors will be able to influence the standards developed to deal with that issue.

4.5 Further developments in the global sphere of authority during the 1980s and 1990s

Besides the renewal of the dimensions of the global sphere of authority, the global sphere of authority has also seen developments in the steering mechanism and compliance dimensions. These developments are not regarded as a policy change, because they do not change the essential characteristics of the global sphere of authority in terms of actors, rules, power, discourse, steering mechanisms and compliance mechanism. Rather these developments have strengthened the already existing steering mechanisms and compliance of the global environmental governance of shipping. Moreover, these developments are important for giving a complete overview of the changing global environmental governance of shipping.

4.5.1 The entry into force of MARPOL Annexes

One of the developments within the steering dimension was the entry into force of the annexes of the MARPOL Convention. The first Annex to enter into force was Annex I, which entered into force when the whole convention did in 1983. Annex I focuses on oil pollution and requires certain construction options, procedures and methods to reduce oil discharges, it imposes discharge limits and it provides the possibility of establishing a special area in which no discharges of oil are allowed except for clean ballast.

The original date for Annex II to enter into force was October 1986, three years after the entry into force of Annex I. However, amendments agreed in 1985 would enter into force only 6 months later, in April 1987, which is why it was decided to postpone the entry into force of the whole Annex to that date. Annex II focuses on the prevention and minimization of operational discharges and accidental releases of chemicals into the sea. Chemicals enter the sea, for example, through the discharge of water that is used to clean tanks. The main chemicals that are carried in bulk are alcohols, vegetable oils, animal fats, petrochemical products and coal tar products (International Maritime Organization, 1998b). The main measures used in this Annex are discharge limits depending on the hazardousness of the substance, carrying a Cargo Record Book and the provision of adequate port reception facilities to receive chemical residues.

The first optional annex of the MARPOL Convention that entered into force was Annex V in December 1988. This Annex focuses on the pollution of garbage by ships. The greatest danger comes from plastics, which can stay into the marine environment for 450 years. Plastics can be ingested by birds, fish and marine mammals. These animals can also become trapped in ropes, nets, bags, etc. with death as a result. When garbage is washed upon the beach, it also pollutes the coastal environment. Annex V prohibits the discharge of plastics and debris, such as synthetic robes and fishing nets, into the sea. It also restricts discharges of other types of garbage (such as food, paper, glass and metal) in coastal waters and special areas and it requires states to provide reception facilities in ports to collect garbage from ships.

The early 1990s saw the entry into force of Annex III of the MARPOL Convention on the prevention of pollution by harmful substances carried by sea in packaged form (July 1992). This annex identified marine pollutants that should be packed and stowed on board of ships in such a way that it minimized accidental pollution. Dumping these harmful substances was prohibited, except in cases where it was necessary for securing safety of ships or life at sea. The ratification and implementation of this annex was hampered by the lack of clarity about how harmful substances carried in packaged form were defined. The already existing International Maritime Dangerous Goods Code was used to provide this clarity in 1991.

The last annex, Annex IV, did not enter into force in the period described here. This Annex on sewage from ships entered into force in September 2003. Since sewage presents health problems and pollution especially in coastal areas, Annex IV sets limits to ships for discharging sewage within 12 nautical miles from land. This Annex also requires adequate port state reception facilities to be present in ports.

4.5.2 Amending the Paris MoU on Port State Control

As explained above, the compliance dimension of the global sphere of authority was changed because of the emergence of a regional MoU on Port State Control in 1982. This MoU enhanced the authority of port states in the global environmental governance while enhancing compliance as well. However, this compliance mechanism was regional and not global. But in 1991, a more global element was added to this specific compliance mechanism.

In that year, the IMO Assembly adopted a resolution on 'Regional Co-operation in the Control of Ships and Discharges'. Through this resolution, the IMO encouraged states to adopt regional agreements on port state control (Keselj, 1999), following the example of the already adopted Paris MoU in 1982 (Hoppe, 2000). Subsequently, the system of port state control demonstrated by the Paris MoU was recognized as a component in the global environmental governance of shipping. Other regions have indeed followed the example of the Paris MoU and adopted mous as well, e.g. the Latin American MoU and the Tokyo MoU were signed in 1993 and the Mediterranean MoU was signed in 1997. Consequently, the regime of port state control has elaborated and turned from a regional compliance mechanism to a compliance mechanism that is much more global.

The early 1990s were also marked by developments within the compliance dimension through amendments to the Paris MoU. In 1992, the Paris MoU on Port State Control was amended to include inspections of operational discharge requirements (Mitchell, 1994). These amendments meant that the crew had to be able to carry out certain shipboard procedures which could be inspected when a ship is in a port. This provision was also adopted in the 1994 amendments to the MARPOL Convention and the four annexes that were in force at that moment. This thus extended Port State Control from dealing primarily with certification, the physical condition of the ship and equipment standards to include operational requirements. With that, the inspection of operational requirements became applicable worldwide instead of only in the European region.

In 1993, a targeting factor was introduced in the Paris MoU at the request of the UK (Tan, 2006), which means that inspections are targeted towards sub-standard ships. The identification of sub-standard ships is based on the profiles of deficient ships and their flag states. Ships are targeted based on their three-year rolling average of inspections and detentions by port states. In the same year, the MoU also started to publish a list of flag states which have a consistently poor safety record. A year later, such a list was also published for single ships. This information and information on inspections, detentions, etc. are published in order to make known worldwide which ships are sub-standard. In a sense, this target factor rewards ships that score well in their inspection and detention rates and puts a disadvantage onto those ships that score above average and are thus more likely to harm the environment.

In 1997, the Paris MoU again introduced some changes in the way in which port state control was organized. Remarkable, these changes were induced by a directive that was adopted within the EU in 1995. The EU adopted EU Directive 95/21/EC on Port State Control because of the increase in the number of substandard shipping trading in the EU (Salvarani, 1996). The reason behind that development was believed to be the incorrect application of port state control despite the existence of the Paris MoU (Salvarani, 1996).

The EU took a two-way approach in strengthening the MoU through this directive. First, the directive implemented the principles of the Paris MoU within the EU, making them mandatory for EU member states. Second, the EU directive included several new provisions that would lead to amendments to the Paris MoU in 1997 (Keselj, 1999; Paris Memorandum of Understanding on Port State Control, 1998). For example, the directive allows the banning of ships from ports and the reimbursement of inspection costs while the Paris MoU did not. (Molenaar, 1996).

Over time, the Paris MoU also expanded geographically. This MoU started with 14 European states, i.e. those states on the West-coast of Europe and the Scandinavian countries. During the 1990s, Canada, Croatia, Poland and the Russian Federation joined the Paris MoU.

4.5.3 New issues and Conventions

The decades 1980s and 1990s do not only show the entry into force of existing measures and the strengthening of Port State Control in Europe, but also the emergence of new issues and measures. One of these issues was the quality of heavy fuel oil and air emissions. This issue was put on the agenda by Norway with the support of the other North Sea states. Besides debating the nature of the problem (to which air pollution problems does shipping contribute?), the nature of the instrument was under debate as well. There were several options, namely adding a new Annex to the MARPOL Convention, amending an existing Annex or adopt a resolution (IJlstra, 1990; Peet, 1991). In 1991, a Resolution, prepared by the MEPC, was adopted by the IMO Assembly on the Prevention of Air Pollution from Ships that called for the preparation for a new Annex to the MARPOL Convention on the prevention of air pollution.

After this decision, the development of Annex VI to the MARPOL Convention took six years until it was adopted in September 1997. Annex VI consists among other of a global cap of 4,5% on the sulphur content of fuel oil and limits to NOx emissions, a ban on emissions of chlorofluorocarbon (CFc) and halons, and the possibility to set special areas (where there is a 1,5% cap on sulphur in fuel oil). While Annex VI was adopted in 1997, it did not enter into force until May 2005.

A second issue that has come up is the environmental impacts of organotin compounds in anti-fouling systems of ships. Anti-fouling paints are used to coat the hull of a ship and to prevent organisms from attaching themselves to the ships hull. Small organisms on a ships hull result in the slowing down of the ship and an increase in fuel consumption. Since the 1970s certain organotin compounds, especially tributyl tin (know as TBT), were used in these anti-fouling paints. Yet, in the mid-1980s, studies in France and the UK showed that marine life is harmed by these organotin compounds that enter the marine environment. The failing of oysters crops in France were in fact the trigger to study this issue. Further research showed that organotin compounds persist in water and kill and harm sea life, such as oysters and whelks.

The first regulatory efforts to address organotin compounds in anti-fouling systems were initially national (for instance France, the US, the UK, Australia and Japan banned the use of TBT) and regional (i.e. the North East Atlantic and EU) in nature. However, they all focused on small ships, i.e. recreational ships less than 25 metres long.

In 1988, during MEPC 26, the Paris Commission of the North East Atlantic requested the IMO to consider the issue and to restrict the use of TBT in anti-fouling systems for ships. In 1990, MEPC adopted a Resolution recommending IMO parties to eliminate the use of TBT on ships less than 25 metres and to eliminate the use of antifouling paints which had a leaching rate of more than 4 mg of TBT per day. The resolution also called for the development of alternatives to TBT based anti-fouling paints and to consider appropriate ways towards a total ban of TBT in anti-fouling paints. In 1998, the MEPC decided to develop mandatory regulations to ban TBT in anti-fouling systems. During this MEPC the call for a global ban of Japan was supported by the North Sea countries (Champ, 2000). The development of this mandatory steering mechanism will be discussed in the next sections when the changes and developments in the late 1990s and 2000s are reviewed.

The last new issue that has come up within the environmental governance of shipping is that of alien species in ballast water. A ship uses sea water to stabilize the ship, especially when it does not carry any or little cargo. This ballast water is discharged when the ship is loading its new cargo. However, ballast water contains organisms and the ship carries these organisms from the place where it loaded the ballast water to the place where it is loading its cargo and releasing the ballast water into the sea or waters of a port. The result is that alien species are discharged in their non-native habitat. In the most extreme cases, this has led to plagues of alien species and to damage to local fish populations and the local ecosystem. It is estimated that ships transfer about 10 billion tons of ballast water every year (De La Fayette, 2001).

The problem of the invasion of alien species through ballast water was already recognized within IMO in 1973 (McConnell, 2002). However, no regulatory initiatives were taken within IMO until Guidelines to combat the introduction of unwanted aquatic organisms were adopted in 1991. Since these guidelines are seen as voluntary, most IMO members did not implement them (North Sea Directorate, 2002). In March 1998, MEPC 42 decided to start developing mandatory regulations on ballast water management (International Maritime Organization, 1998b). The development of this mandatory steering mechanism will also be discussed in a later section.

4.6 The authority of the state during the 1980s and early 1990s

The previous sections have shown that the global sphere of authority has changed considerably during the 1970s and the beginning of the 1980s, while it remained stable during the 1980s and mid-1990s. This section will focus on the way in which these changes have affected the authority of the state, both in developing steering mechanism as in generating compliance with steering mechanisms.

4.6.1 Authority in developing steering mechanisms

As described above, one of the differences compared to the situation in the 1950s and 1960s is the opening up of the global sphere of authority to port and coastal states. In the 1950s and 1960s, IMO (IMCO) was mainly dominated by the largest flag states, both as members within IMCO's institutions as in the negotiations. It were also those flag states experiencing pollution (i.e. because they are also coastal states) that gave the impetus to develop international standards for oil pollution from shipping. However, the institutionalization of port and coastal states and the increased participation of such states have affected the primacy of flag states in IMO. Flag states are no longer the only actors that have the authority to develop steering mechanisms for ships, because port states have this authority as well. Formal rules of the game, i.e. UNCLOS, have turned the jurisdiction to set standards over ships into a shared jurisdiction.

One of the consequences is that the interdependence between the state members of IMO has changed. To recall, the interdependence was mainly between flag states and was based on the fact that those flag states experiencing pollution could not set standards over foreign ships that are passing through waters close to their shores. The interdependence now depends on the situation. For example, a port state is now less dependent on flag states, because it is able to set standards over ships entering its port. A coastal state is, however, still dependent on the flag state, because a coastal state is still not able to set such standards. Of course a port is often a coastal state as well. This means that the port state can only set standards for ships that actually visit a port and not those that only pass through the waters. Thus the interdependence is still there and is even more widespread, because of the increase in actors that are able to set standards that affect foreign ships and that participate in the global sphere of authority. Still, port states have gained a partial way out because of their competence to set standards over ships.

However, port states will not easily favour this option, because of competitive disadvantage that they can create compared to other ports in neighbouring countries. This is also why the dominant discourse is still that global standards are the best way to set standards for shipping, also with relation to port states. As explained above, the call and the discourse for global standards still put a lot of pressure on both flag and port states to not threaten the level playing field by adopting unilateral or regional standards.

Thus, it can be concluded that the rationale for developing global steering mechanisms has remained the same, even though a new type of actor has entered the global sphere of authority. The next question is whether the policy change has affected the authority and influence that flag states have over the development of steering mechanisms.

Besides a growth in the number of state actors involved in the global sphere of authority, there has also been a growth in the number of observers. Imco knew 11 observers in 1966, 21 in 1971 and 45 in 1978 (M'Gonigle & Zacher, 1979). The majority of these observers are industry associations, but environmental NGOs have become part of the actor dimension as well, e.g. Friends of the Earth International received observer status in 1973.

It was noted that in the 1950s and 1960s, most of the industry associations and NGOs were not directly involved in the negotiations within IMCO. However, this changed in the course of time for the industry associations. For the industry, the underlying reason was probably that the industry was more and more under pressure to decrease its pollution and regulations proposed would set costly requirements. By the end of the 1970s, the International Chamber for Shipping and the Oil Companies' International Marine Forum had become extremely important lobbyists on environmental matters at IMCO (M'Gonigle & Zacher, 1979).

Furthermore, their authority and influence was based on the fact that the industry controls much of the information and the technology necessary for pollution control (M'Gonigle & Zacher, 1979). For example, it was again the oil industry that after already developing the Load on Top method developed the Crude Oil Washing method. The development of this relatively cheap method was spurred by threats of adopting requirements that would be more expensive. In that sense, the oil industry had a considerable influence on the 1978 conference at which the Protocol to the MARPOL Convention was adopted, because Crude Oil Washing was accepted as a viable alternative to the Load on Top method.

The oil and shipping industry also had considerable influence on Annex VI of the MARPOL Convention. The initial proposal was to establish a global cap of 1,5% on sulphur in fuel oil. A 1,5% sulphur cap effectively means that ships would have to switch to distillate fuels that have much lower sulphur contents and thus much less so_2 emissions.

Within the negotiations on the establishment of a global sulphur cap, and especially the specific limit imposed, three coalitions emerged. First, the 1,5% proposal was rejected by the oil industry and the states of the Organization of the Petroleum Exporting Countries (OPEC) that were fearful for their markets in heavy fuel (Kütting & Gauci, 1996). The oil producing OPEC countries would have to invest in desulphurization plants, which would result in higher production costs (Tan, 2006). The other coalition consisted of states that have many ships under their flag, e.g. Bahamas, Bahrain, China, Egypt, Greece and Liberia proposed a cap of 4 to 5% (Kütting & Gauci, 1996). These states with maritime interests together with the shipping industry opposed the higher costs that would be related to buying low-sulphur fuel (Tan, 2006). The last coalition were those states that are affected by acid rain, e.g. Sweden, Norway, Denmark, Finland, Germany, the Netherlands, but also Japan and Korea (Interview International Maritime Organization, 2007; Kütting & Gauci, 1996). Yet it should be noted that the North Sea states interests were not only environmental, because they would also benefit economically from a stringent cap as they produce low sulphur fuel oil themselves (Tan, 2006).

To accommodate the higher expectations and need of certain states, the possibility to designate special areas in which the maximum sulphur content would be 1,5% was proposed (Kütting & Gauci, 1996) and pursued by the states bordering the Baltic Sea (Tan, 2006). The proposal was supported by the oil industry to stave of demands for a stricter global cap (Tan, 2006). Still, the European states kept insisting on imposing a global cap as well. In the end, due to powerful objections of the oil industry it was decided to impose only a 4,5% cap, even though the internationally agreed maximum sulphur content in marine fuel at that moment was already 5% (Ninaber, 1997; Tan, 2006).

It should be mentioned that much of the authority that the industry has in shaping steering mechanisms is through states that share their interests. Coalitions are not only built between states, but also between states and industries, i.e. in the case of air pollution the OPEC states with the oil industry and flag states with the shipping industry. It is this combination that makes the industry voices so powerful.

Besides the flag, port and coastal states and the industry, environmental NGOs were part of the global sphere of authority. Moreover, they are regarded by governmental delegates as having influence over decisions made within IMO (Peet, 1994). The main issues where the efforts of environmental NGOs have been successful were the issue of Particularly Sensitive Areas, the negotiations on Annex III of MARPOL and the North Sea Special Area (Peet, 1994) and more recently the issue of ship recycling (Interview Department for Transport, 2007). According to FOEI (Interview Friends of the Earth International, 2007), the strategy that seems to be most effective is to use arguments rather than obstruction. The environmental NGO that has a more obstructive approach, Greenpeace, is struggling to keep its position within IMO, because it is regularly breaking IMO rules itself (Interviews Department for Transport, 2007; International Maritime Organization, 2007; Stolt-Nielsen Transportation Group, 2007). The environmental NGOs thus have gained authority over the development of steering mechanisms. Yet, among others, because they are limited in numbers, their authority is smaller than that of industry observers (Interview International Maritime Organization, 2007; Peet, 1994).

All in all, the power struggle between actors having environmental concerns and advocating stringent environmental standards versus those actors that want as less regulation as possible, which also existed during the 1950s and 1960s, has not changed much. It is still about the struggle to find a compromise between the environmentalists' and the maritime interests. However, the actors and the coalitions behind each of these interests have changed. According to Tan (2006, p. 102), 'the fundamental difference lies in the fact that the commercial shipping interests do not presently enjoy the full and unwavering support of their traditional maritime state allies'. This is partly related to the fact that the centre of gravity of the shipping industry has moved from the traditional maritime states towards the developing countries.

Thus, while in the early years of IMCO the maritime states by and large advocated their maritime interests, except for when they had clear environmental problems along their coast. The industry associations both pressured these flag states to do so and supported these flag states during international negotiations. However, currently, some of these flag states have lost parts of their fleet, changing the emphasis towards their port and coastal state activities and environmental interests. Other states, namely those states with open registries, have taken the place of the traditional maritime countries. Thus, traditional parties have become more active participants on the environmental side, because they are now able to voice their coastal concerns within IMO (e.g. Australia and Canada, but also the UK, the US, the Netherlands, etc.). At the same time new states have become part of IMO, because their fleet has expanded over the years (e.g. Panama, the Bahamas, China, Cyprus, Malta and Singapore).

4.6.2 Authority in generating compliance

While the enforcement and compliance with global standards was very weak during the 1950s and 1960s, we have seen that new compliance mechanisms have become part of this global sphere of authority. These new compliance mechanisms are part of the MARPOL Convention or adopted through amendments within the Paris MoU on Port State Control. This means that they are either part of the steering mechanism that sets out the global standards, or that they are an initiative of port states who seek better control over those ships entering their ports.

The first change in the compliance dimension that influences the way in which states can ensure compliance is the requirement stated within MARPOL Annex I that ships have to carry an International Oil Pollution Prevention certificate. This certificate shows whether a ship complies with the equipment standards of MARPOL, i.e. Segregated Ballast Tanks and equipment related to Crude Oil Washing, and is valid for five years. The flag state is responsible for issuing and updating these certificates. With that the flag states finally gained a compliance mechanism that increased their authority in generating compliance.

However, most flag states ask classification societies to do the required surveys to judge whether a ships complies with MARPOL Annex I or not and is eligible for receiving the certificate (Interview Lloyd's Register, 2007; Vorbach, 2001). It should be noted that surveys and the involvement from classification societies was already common within the shipping industry, however not in terms of environmental standards. Classification societies have always been involved during the construction of a ship and check whether a ship complies with safety standards (Furger, 1997). Classification societies develop rules and standards for ship construction. When a ship is built according to these rules and standards it receives certificate of classification of the classification society (Furger, 1997; Interview Lloyd's Register, 2007). However, the requirement within MARPOL to carry certificates requires surveys to assess whether ships comply with Annex I standards. Because of the lack of technical expertise with the government, a lot of flag states have delegated this responsibility to classification societies, who do have the required technical expertise (Interview Lloyd's Register, 2007). In other words, classification societies therefore have gained authority in ensuring compliance in the environmental governance of shipping.

In addition, the MARPOL Convention granted port states the right to inspect whether the certificate matches the equipment onboard. Ships that do not meet the equipment standards are detained in port until they do comply. According to Mitchell (1994), since this detention only affects the ship itself and not the flag states, it was not perceived as a potential threat to flag states' sovereignty (unlike prosecutions and fines in the past).

All in all, since the number of discrepancies between the International Oil Pollution Prevention certificate and onboard equipment declined in the years after MARPOL came into force, one can conclude that this way of enforcing and ensuring compliance with oil standards was much more effective than the discharge limits and port state inspection of the OILPOL Convention (Mitchell, 1994). Undoubtedly, the coordinated effort of port state inspection within the Paris MoU facilitated this. The move away from discharge limits (although they still applied) to equipment standards made it easier for both flag and port states to detect non-compliance (Mitchell, 1994).

Moreover, over the years, the other Annexes of MARPOL entered into force, elaborating pollution standards to other discharges and emissions. The way in which enforcement is arranged and compliance is ensured depends on whether discharge and/or equipment requirements are at the heart of each of the Annexes. Discharge requirements are monitored using Record Books, something that we saw in the OILPOL and still exists within MARPOL Annex I. It will be no surprise that ensuring compliance with discharge limits was still very difficult. The only mechanisms through which compliance can be controlled are through the Record Books and Port State Control. However, both mechanisms are easy to bypass. Compliance with equipment standards, on the other hand, is controlled through a five-year certificate that states that the standards are met, just like with the International Oil Pollution Prevention certificate. Flag states, via their classification societies, and port states are therefore much better able to ensure compliance.

Most annexes impose both discharge and equipment standards onto ships. The only exception is Annex V on garbage. This Annex solely relies on the Garbage Logbook and Garbage Record Book to control compliance. It can therefore be no surprise that compliance with Annex V is very low. The other annexes rely on the combination between the record book and a certificate, e.g. the Cargo Record Book Prevention Certificate for the Carriage of Noxious Liquid Substance in Bulk for Annex II or the International Air Pollution Prevention certificate for Annex VI. In these cases, the compliance with the equipment standards is ensured through the certificate which is controlled by both flag and port states. Still, the compliance with the discharge limits is not ensured because of the difficulties just mentioned.

Besides the increased authority of both flag and port states in enforcing global standards, another development is noteworthy. Over the years, with the entry into force of the other Annexes, more certificates proving compliance within environmental standards have come into existence. This also means that the authority of classification societies in controlling compliance has increased. After all, most flag states have given the responsibility for issuing certificates to classification societies. In other words, ensuring compliance, although formally in hands of flag states, is to some extent becoming a non-state matter (Vorbach, 2001).

Besides these certificates, port state inspections have also become one of the main mechanisms through which compliance is ensured (Interview International Maritime Organization, 2007; Mitchell, 1994; Vorbach, 2001). This increased authority of port states is not only related to provisions within MARPOL which have expanded the rights of port states in inspecting ships and detaining them. It is also related to the Paris MoU and changes within this MoU.

In the 1980s, when the MoU had just come into being, port state control focused on equipment standards. This is very much related to the fact that discharge requirements are impossible to detect during a port state inspection, while the equipment requirements and the associated certificates are a good target for port state inspections. The Paris MoU coordinated port state inspections between its 14 members and set the objective of inspecting 25% of incoming ships.

Three important changes enhanced the strength and focus of these inspections. First, operational requirements were included in the inspections as of 1992. This was thus a direct expansion of the authority of the port state in controlling compliance. However, one can question what this change exactly means for the authority of port states over compliance, because controlling operational requirements while in port does not mean that the crew conducts the operations while at sea. Second, a targeting factor was adopted, resulting in increased inspections of sub-standards ships. Finally, within the European Union, port state control has been further strengthened by the adoption of the Directive on Port State Control in 1995. European member states are now legally required to implement provisions of the Paris MoU.

In the 1950s and 1960s, one of the weaknesses of the port state inspection regime was the lack of sanctions that port states can impose on ships. The follow up of inspections, i.e. in the form of prosecutions, was the responsibility of the flag states. During the 1970s and 1980s, however, port states were granted two ways in which they can follow up inspections. They can detain a ship in port when a ship is not in compliance or prosecute it when it has violated discharge limits somewhere at sea. The compliance mechanism of port state control has therefore become stronger and has become a compliance mechanism that is detached from the flag state. The interdependence between port states and flag states in terms of generating compliance is therefore much less than it was in the 1950s and 1960s.

The final compliance mechanism in this global sphere of authority is reporting. Most environmental treaties require parties to report on the implementation of the treaty to be able to evaluate and improve compliance. To recall, there are three aspects that parties to IM(C)O have to report on: available reception facilities, inadequate reception facilities, enforcement.

Mitchell (1994) has researched the reporting trends on all three aspects. Reporting on available reception facilities has mainly been done through surveys organized by the IMO secretariat, but overall the response was limited (around 30%). The reporting on inadequate reception facilities was also done through surveys, but response to these surveys was even lower than to the available reception facilities surveys. Moreover, the shipping industry, instead of states, provides information on inadequate reception facilities. Yet, according to IMO (Interview International Maritime Organization, 2007), they are reluctant to do so, because it could backfire on them.

Finally, Mitchell (1994) also notes that with regard to reporting on enforcement, there was increased attention to reporting on enforcement in the early 1970s because of the MARPOL Convention and because of recommendations from MEPC. This has led to an increase in reporting, from a 6% response rate in 1975 to a 49% response rate in 1990. Furthermore, the quality and content of these reports varied immensely across countries, with even more than 30 countries that never submitted a report. All in all, reporting is a compliance mechanism that looks very promising on paper, while it is very unlikely that reporting will strongly enhance compliance in practice. This means that the authority of IMO over compliance remains limited.

It can be concluded that the authority of the state in generating compliance has increased at the expense of the authority of the industry. However, it depends on which type of state you are referring to, because it is especially the authority of port states that has increased. An extensive system of port state control has emerged that puts more pressure on the industry to comply. Flag states still have limited ways to enforce environmental standards, although they have gained an extra compliance mechanism in the form of certificates. Yet generally, these certificates are handled by the classification societies.

4.7 The emergence of an European sphere of authority during the 1990s and 2000s

One of the most recent changes in the environmental governance of shipping is the emergence of the European Union as a sphere of authority for shipping. The EU started to focus on shipping and to develop separate steering mechanisms in the early 1990s. This section will explore why a European sphere of authority emerged and what the main results are of this sphere of authority.

4.7.1 European Directives and the European Maritime Safety Agency

In 1993, the European Commission (EC) published a Communication on A Common Policy on Safe Seas. This was a reaction to the realization that the number of substandard ships (i.e. those ships not adhering to the global standards of the IMO) trading with the EU had dramatically increased in the early 1990s (Salvarani, 1996). The European sphere of authority that emerged subsequently focused on eliminating substandard shipping from EU waters. Hence the focus is not only on the environment, but also on safety. Several European steering mechanisms, i.e. directives and regulations, followed on issues like a notification system for ships carrying dangerous or polluting goods; on training, recruitment and working conditions of seafarers; marine equipment, etc. Other non-environmental issues for which EU legislation was developed over the years are for example: classification societies; port infrastructure; monitoring, control and information system for maritime traffic; and liability and compensation for oil pollution. According to Nollkaemper and Hey (1995), the EU legislation adopted after 1993 consist of substantive measures with the objective of laying down regional standards for shipping activities instead of merely a means to implement existing international regulations. Likewise, Urrutia (2006) explains that even though some legislative decisions were made between 1978 and 1992, the starting point of the EU maritime safety policy is the 1993 Communication on A Common Policy on Safe Seas.

Especially important for the environmental governance of shipping, was the believe that Port State Control under the Paris MoU was applied inconsistently across the Paris MoU members (Salvarani, 1996). This sparked the desire of the EC to develop European legislation on Port State Control. In 1995, the EU Directive on Port State Control was adopted. This Directive had a direct effect on the Paris MoU on Port State Control. More specifically, there are two ways in which this directive strengthened the Paris MoU on Port State Control (Keselj, 1999). First, because it implements the principles of the Paris MoU within the EU. While the Paris MoU depends on political will, the EU directive is mandatory for EU member states. Second, the EC directive included several new provisions that have stimulated amendments to the Paris MoU in 1997 (Keselj, 1999; Paris Memorandum of Understanding on Port State Control, 1998). For example, the directive allows the banning of ships from ports and the reimbursement of inspection costs (Molenaar, 1996).

The development of a European sphere of authority received a new political impetus after the tanker accidents of the Erika and Prestige (Interview European Commission, 2007a; Interview Ministry of Transport, Public Works and Water Management, 2005a; Urrutia, 2006). The single hull tanker Erika broke in two near the coast of France in 1999. More than 10,000 tons of oil was spilled and between 120,000 and 300,000 birds were killed. The accident with the Prestige happened in November 2002. This single hull tanker sank of the coast of Spain spilling 77,000 tons of oil, contaminating beaches in Portugal, Spain and France, and killing 65,000 to 130,000 birds. Both accidents elicited initiatives to develop European legislation to prevent oil pollution that would go further than the existing IMO standards.

Both after Erika and after the Prestige accident, the EC responded with proposals for EU legislation that would accelerate the phasing out of single hull oil tankers and the implementation of double hulls on oil tankers. This EU legislation would not only apply to ships registered in EU member states, but also to foreign ships that would visit a European port. Such a phase out schedule for single hull tankers was already adopted by the US unilaterally in 1990 and by the IMO globally in 1992, but the EU wanted to fasten the adopted global schedule. In both cases, the initiatives of the EC and the EU member states marked the start of new debates on this issue within IMO.

After the Erika accident, the EC proposal was published in March 2000. This initiative also led to debates on this issue within IMO. Since 'unilateral' action by the EU would threaten the existence of the uniform standards set by the IMO, the Secretary-General of the IMO changed the dates of the next MEPC meeting to as early as possible and took every opportunity to insist that regulations should be developed within IMO (De La Fayette, 2001). In April 2001, the MEPC indeed adopted a new and faster schedule for the phase out of single hull tankers. Still, these amendments were not fully in line with the deadlines that were proposed by the EC (Tan, 2006). The EU followed up the adopted global schedule with the adoption of EU regulations in February 2002. These regulations were in line with the deadlines adopted within IMO and were hence a watered down version of the initial proposal of the EC.

After the Prestige accident, the sequence of events was slightly different. This time the EU adopted its own schedule without waiting for the IMO (Urrutia, 2006). The EC proposal for again a new and faster schedule for the phase-out of single hull tankers was published in December 2002. In July 2003, the EU adopted the regulation that brought this schedule into force. In the same month, an MEPC meeting was held in which the delegation of Italy, as spokesperson for all 14 EU member states, proposed to bring the global standards inline with the EU legislation on the deadlines for the phasing out of single hull tankers (Marine Environment Protection Committee, 2003). The MEPC responded by adopting and copying the EU legislation on the global level in December 2003.

Next to developing European legislation to eliminate substandard shipping, the European Maritime Safety Agency (EMSA) was established in June 2002. Just like the development of the EU legislation on oil pollution from shipping, EMSA was also a reaction to the Erika and Prestige accidents. Emsa provides technical and scientific advice to the EC in the field of maritime safety and prevention of pollution by ships, especially in updating and developing new legislation. Moreover, EMSA monitors the implementation and evaluates the effectiveness of EU regulations (Interview European Commission, 2007a). The environmental issues EMSA specifically focuses on are the provision of port reception facilities and the prevention of air pollution from shipping. EMSA is furthermore asked to track the developments within IMO of the newest issues that have been raised within MEPC. Since it is also active in Port State Control and the Directive on criminal sanctions, EMSA also indirectly plays a role in the enforcement of international environmental regulations.

Besides the more substantial issues that European governance of shipping covers, one recent development has been the adoption of the Directive on ship-source pollution and criminal penalties in 2005. This directive aims to improve maritime safety and the protection of the marine environment from pollution by ships by ensuring that persons responsible for discharges are subject to adequate penalties. This Directive makes penalties for marine pollution compulsory, however, the individual member states and not the EU as a whole are the ones that have to decide on the type and level of these penalties. The shipping industry and the three major EU flag states (Greece, Malta and Cyprus) are far from happy with this directive. Intertanko and some other industry groups even have taken the EC to the English High Court arguing that the directive is in conflict with certain provisions of the MARPOL Convention. The English High Court referred the case to the European Court of Justice. The European Court of Justice ruled that the directive was valid and that since the EU is not a Party to the MARPOL Convention it is not bound by this Convention.

In 2005, the European Commission presented a set of 7 new directives and regulations (Urrutia, 2006). In March 2009, the European Parliament adopted this set of European legislation. Two directives are relevant for the environmental governance of shipping: a new directive on port state control and a directive on compliance with flag state requirements. The first indicates new amendments to the Port State system in Europe, namely the replacement of the target for individual EU member countries to check 25% of ships by a collective target for Europe as a whole to check all ships (European Commission, 2009). Moreover, more frequent inspections have to be made of high-risk ships. The directive on compliance with flag state requirements aims to improve flag state implementation and compliance within the EU.

These developments show that, in the last 15 years, the EU has been developing a separate sphere of authority from the already existing global one. According to Roe (2009), this development is a reaction to the failure of the traditional and more hierarchical global sphere of authority, i.e. IMO, to adequately reflect the activities, desires and ambitions of its stakeholder. However, there are also several examples of EU legislation that only serve as a means for implementing IMO Conventions within the EU. In such cases, the EU follows the work of IMO and pursues European legislation only to enhance and harmonize the implementation of IMO conventions between the EU member states. The question is how the activities of the EU in governing the environmental impacts of shipping and the emerging European sphere of authority conflicts with, interacts with and changes the global sphere of authority.

4.7.2 *The EU sphere of authority and changes within IMO*

The first change in the global sphere of authority that occurred as a result of the emerged EU sphere of authority is related to the 1995 Directive on Port State Control. The influence of this EU directive in the global sphere of authority lies in the changes it has caused within the Paris MoU on Port State Control. The EU directive makes Port State Control mandatory for EU member states. Furthermore, the existence of the directive gives the European Commission the possibility to start the infringement procedure against those member states that fail to implement and/or carry out the requirements of the directive (Interview Ministry of Transport, Public Works and Water Management, 2005a; Molenaar, 1996). In other words, the EC has a compliance mechanism that ensures that EU port states carry out port state inspections, which in turn affects all ships – foreign or not – that visit a European port.

What thus happened with the adoption of this directive was that the European institutions and the EU member states agreed to take enforcement and compliance matters into their own hands. This European steering mechanism strengthened compliance with IMO standards, because it would ensure proper implementation of the Paris MoU and it strengthened Port State Control by bringing about amendments to the Paris MoU. Yet, the fact that the directive focused solely on Port State Control keeps the effect of this directive on the global sphere of authority limited to the compliance dimension.

In contrast to the Directive on Port State Control, the policy practices of the EU focusing on the prevention of oil pollution from shipping in the early 2000s have caused changes in all dimensions of the global sphere of authority in environmental governance of shipping. The EU launched an initiative to accelerate the phasing out of single hull tankers to times, after the accident with the Erika and after the accident with the Prestige.

What was very important in both cases was that the actors of the existing global sphere of authority were concerned that the European steering mechanisms would compete with the IMO steering mechanisms and global sphere of authority. As Frank (2005) explains, during the negotiations of this issue, the proposal of the EU caused some concern among IMO members about the repercussions that unilateral EU legislation would have on global shipping and the security of oil supplies. IMO members were particularly disturbed by the decision of the EU to, after the Prestige accident, adopt the legislation without waiting for the outcome of the negotiations within IMO. Indeed, as Blanco-Bázan (2004, p. 283) observes 'after the Erika it apparently became acceptable to press IMO with either amendments to MARPOL to satisfy a group of countries or unilateral action coming from them'. Moreover, according to the International Chamber of Shipping, the EU's actions generated discomfort within the IMO, because the EU politicized debates that used to be very technical in nature (International Chamber of Shipping & International Shipping Federation, 2006).

The reason behind these concerns is that the unilateral action of the EU was considered to challenge the uniformity of current shipping regulations and the authority of the IMO and would set a precedence for other regions to adopt their own regulations (Frank, 2005). What thus happened was that the steering mechanisms of the EU entering the global sphere of authority were in conflict with the existing discourse that emphasized global standards as the only way through which a level playing field can be ensured. This discourse around global regulations and the IMO being the appropriate body to regulate shipping had already been dominant for several decades. The EU steering mechanism created enormous pressure on especially those actors that have maritime interests to make sure that this discourse stayed prominent. In the case of the Erika accident, they had to accommodate the concerns of the EU in such a way that the EU would not adopt a steering mechanism which would establish diverging European standards. However, after the Prestige incident, these maritime interests were put under even more pressure because the EU had already adopted its own standards. Subsequently, the only way to keep the standards uniform and the discourse alive was to adopt similar, global requirements as those of the already adopted EU regulations.

Disputing the dominant discourse thus also affected the power dimension within the global sphere of authority. The power relations between the two dominant interests were affected, because environmental concerns gained more weight than maritime concerns. Yet, another way in which the power relations were affected was via the rules of the game and the actor dimensions. Because of the adopted EU regulation, the EU gained so-called external competence on the issue of phasing out single hull tankers. In the EU, there is internal competence on those issues that are subject to European legislation. Furthermore, the rule is that when the EU has internal competence on an issue, it also has external competence on that issue. This external competence means that the EU member states have to act as a coalition and speak as one voice within international negotiations on that issue.

Within shipping, the fact that there was now EU legislation on single hull tankers meant that the previously individually operating EU member states were formally required to go to the MEPC meetings with a coordinated position. In other words, new rules of the game started to prevail which resulted in the EU acting as a coalition with one position in the global environmental governance of shipping. According to the European Commission (Interview European Commission, 2007a), the fact that this was a new situation meant that the EU member states had to get used to their coordinated position and would sometimes voice another position than that of the EU as a whole. Moreover, according to a civil servant of the UK (Interview Department for Transport, 2007), some IMO members are suspicious about what the EU wants within IMO and have concerns about the role that the EU plays within negotiations, because it tends to politicize a debate that should be technical in nature. In addition, as the European Commission confirms (Interview European Commission, 2007a), the EU is not always recognized as a coalition, because the EU is not a flag state, and because some members of the IMO did not know the EU and how the EU functions.

Still, what followed was that these new rules of the game brought about a new coalition, i.e. of the EC together with the EU member states, within the negotiations on the single hull issue. And indeed, during the MEPC meeting of July 2003, the delegation of Italy was the spokesperson for the other 14 EU member states proposing to bring the global schedule in line with the EU one. This coalition of the EC and the EU member states also brought more power to the diverging phase-out schedule and the environmental concerns that the EU promoted.

4.8 The EU sphere of authority and the renewed global sphere of authority during the 2000s

As argued above, important in this policy change has been the proposals for and the adoption of EU steering mechanisms that diverge from the existing global ones. Such proposals for and the adoption of steering mechanisms within the EU resulted in a new dynamic within the global sphere of authority, causing the renewal of the global sphere of authority. However, the EU sphere of authority together with the renewed global sphere of authority only matter when the EU is either deliberating new standards or has already adopted these new standards within the EU. Section 8.1 will discuss how the EU sphere of authority affects the renewed global sphere of authority in such cases. In other issues, however, the global sphere of authority as it existed since the 1980s continues to govern the marine pollution from shipping! This will be discussed in Section 4.8.2. Finally, Section 4.8.3 will explore a change in the global sphere of authority that has occurred in recent years.

4.8.1 Diverging European standards and IMO

It is important to note that only when the EU is considering adopting diverging EU standards for shipping, EU legislation starts to influence and change the global sphere of authority. Then pressure is put on the global sphere of authority and the results are new dynamics in policy making practices. The directive on Port State Control and the regulation on phasing out single hull tankers are not the only examples in this regard. Another important example lies in the area of air pollution.

The IMO is currently finalizing the revision of Annex VI of the MARPOL Convention which deals with sox and nox emissions from shipping. Annex VI was adopted in 1997 and entered into force in 2005. Immediately after its entry into force, it was decided to start the revision of the Annex among others because compared to the contribution of land-based air emission the contribution of ships' air emissions to local air quality problems is growing and new technologies have been developed (Interview International Maritime Organization, 2007).

Since 1999, the EU developed its own legislation and a strategy targeting among others sulphur emissions from fuels used by ships in European ports. The most recent Directive, adopted in 2005, implements the provisions of Annex VI of the MARPOL Convention, but adds a new requirement. Ships at berth in an EU port have to use fuels with only 0,1% sulphur as of January 2010. However, remarkable, this difference between the global and EU standards has not been an issue within IMO. According to the European Commission (Interview European Commission, 2007b), this might related to the fact that the fuel used in ports is only a few per cent of the total fuel used by ships and because similar regulations already exist in the US. It does, however, show the willingness of the EU to do something about the sulphur emissions coming from shipping, an issue that is directly related to Annex VI. This willingness stems from the fact that in the EU land-based transport is already legislated and is reducing, while ships emissions were not yet regulated and are expected to rise considerably in the coming decade (Interviews European Community Shipowners' Association & European Sea Ports Organisation, 2007; Transport & Environment, 2007).

The EU has therefore not been silent during the negotiations of the revisions for Annex VI. However, this time, the EU did not use existing EU legislation, but the threat of future EU legislation to pressure the MEPC to adopt stringent standards for sox emissions. If the adopted global standards would not be satisfactory, the EC would use the revision of the 2005 European directive on sulphur to propose new EU standards on SO_x emissions (Interview European Commission, 2007b). And indeed, the EU seems to get its way again, because the MEPC agreed to a considerable restriction of sulphur content of fuels in the future in April 2008.

Thus in cases where the EU as a community of port states (threatens to) develops steering mechanism with requirements that apply to foreign ships as well, the global sphere of authority no longer has the monopoly on developing standards for shipping. In those cases, an EU sphere of authority emerges that challenges and complements the global one.

The emergence of an EU sphere of authority in turn challenges the existing discourses in the global sphere of authority of a level playing field for the shipping industry, the necessity of global standards and the IMO as the competent body to develop these global standards. Even though these discourses are challenged, they remain dominant in the renewed global sphere of authority. First, because EU legislation is not seen as a legitimate alternative by the wider shipping community and most members of IMO. Second, because the EU member states also still regard it most desirable if the IMO develops global standards and the level playing field for shipping is ensured. For them EU legislation is the last resort. One discourse that is however different is the technical nature of the issues and debates within IMO. The EU has often been accused of politicizing the debate (Interview Department for Transport, 2007).

The development of EU steering mechanisms also has consequences for the actor dimension of the renewed global sphere of authority. Because of the European system of internal and external competence, the EU member states have to coordinate their position within IMO (external competence), when EU legislation on that issue exists (internal competence). When this EU legislation does not exist, but there are future plans to do so (like in the air pollution issue), there is more freedom for EU member states to engage in a coalition or not. Still, the threat of future EU legislation does give the EU a sense of unity in such issues at least in the eyes of all the other actors in the global sphere of authority.

Another development in the actor dimension is the increased effort of the EC to become a more recognized actor in the global sphere of authority. The EC is an observer, but next to that appointed a permanent representative at the IMO in London two years ago. This person follows all kinds of meetings and coordinates the input of the EC in the IMO (Interview European Commission, 2007a). The EC is thus much more visible now than it was 10 to 15 years ago, when the EC was an observer and operated much more independently from all the individual EU member states.

The power dimension of the global sphere of authority is renewed as well. The European coalition brings a considerable weight to the environmental side and subsequently changes the balance of power that existed within the global sphere of authority. One of the ways in which the EU coalition brings about this change in the power relations is through challenging the existing discourse on global standards and level playing field which exists in the global sphere of authority. In cases where the EU has already taken unilateral action by developing European legislation, the EU creates an impetus to smoothen the distortion of the level playing field on the global level. The only way to do that is to bring global standards in line with EU standards. In addition, in cases where the EU threatens with the development of EU legislation, maritime actors are taking efforts to look for a way to accommodate the EU concerns with global standards that are acceptable to the shipping industry as well. According to Tan (2006), there are two reasons why the EU threats to pursue unilateral legislation are this powerful. First, because many ship owners depend on trade with the EU. Second, because the shipping community is aware of the power and inclination of the EC to enact legislation onto the EU member states.

Moreover, in recent years there has been another development that has enhanced the position of the EU in cases where the EU as a whole actively seeks to influence the steering mechanisms developed. That is that in 2004, 10 new countries joined the EU. Among those 10 countries are the two large flag states of Malta and Cyprus. Now that these two countries are part of the EU, the EU's registered tonnage comprises about 25% of the world tonnage (International Chamber of Shipping & International Shipping Federation, 2006). This adds considerably to the recognition and power of the EU in the global sphere of authority.

Finally, the renewed sphere of authority also shows a difference in its steering mechanisms dimension. The EU steering mechanisms become part of the global sphere of authority, because of their impact on foreign ships. The result of the emergence of the EU sphere of authority is thus that stricter standards become part of the global sphere of authority.

The conclusion is that the global sphere of authority is no longer a steady sphere of authority across all environmental issues negotiated, because the global environmental governance of shipping is no longer characterized by a steady and uniform global sphere of authority. Granted, the global sphere of authority is still centred on the IMO with its flag and port states and with vested maritime interests that are under pressure from environmental interests. Yet, the EU has brought about a new, albeit wavering, dynamic within that sphere of authority.

It is a wavering dynamic because the EU sphere of authority emerges only on those issues where the EU believes that there is a policy gap that needs to be filled. As also observed by Pallis (2006), EU member states, the EU institutions and the public are sometimes unhappy with the lack of progress within IMO or the poor effectiveness of existing global regulations and the consequences that this may have on either the shipping industry, the seafarers or the environment. This in the end diluted the reluctance of member states to develop regional measures. This reluctance existed because until today, consistent with the discourse within the global sphere of authority, the general preference within the EU is to develop regulations for shipping on the international level (Interview Ministry of Transport, Public Works and Water Management, 2008). However, in some cases, the necessity to regulate shipping was so real for the EU member states and the EU institutions that they decided to develop EU legislation to deal with an issue.

As we have seen, this was especially the case in the oil pollution issue in the early 2000s and to a lesser extent with sulphur emissions from shipping more recently. The two major oil spills from Erika and Prestige in European waters turned the attention of both the public and the policy makers to the risk of oil pollution from shipping within the EU. Enough reason for the EC to propose EU legislation on oil pollution. In addition, the EU sulphur emissions from shipping are expected to pass those of road transport in the near future. This has also aroused attention from environmental interest groups, the shipping industry and the policy makers. A reason for the EU to increase pressure within IMO to deal with this issue more effectively. These specific European activities in turn have informed the decision making process within IMO.

4.8.2 The continuing conventional IMO and the EU

However, it should again be stressed that the EU sphere of authority and renewed global sphere of authority only exist in cases where EU (threatens to) develop(s) diverging standards. In cases where the EU is not taking a regional initiative for new steering mechanisms, the global sphere of authority still has the monopoly in the global environmental governance of shipping. Furthermore, in those issues, the global sphere of authority is depicted by its more traditional characteristics and interactions between maritime actors and environmental actors as described in Section 4.4.

Moreover, most of the time there is no EU legislation on an environmental issue or the EU legislation that does exist is in line with the global standards. Eu legislation then simply serves as a means for implementing the global standards within the EU. In these cases, the EU legislation is not part of the global sphere of authority, but can be considered as an internal measure within the EU.

The first issue in which this is the case is port reception facilities. MARPOL requires adequate provision of port reception facilities to deal with waste for which discharge limits exists, i.e. oily waste, sewage, household waste, chemical waste, etc. The provision of adequate port reception facilities by ports has already been problematic since the 1950s. At the end of the 1990s, the European Commission proposed to develop a directive on Port Reception Facilities to implement and enforce this IMO requirement within the EU. The Directive was adopted in 2000 and has basically two requirements for ports and one for ships: (i) ports have to develop a port waste management plan, ii) ships have to notify a port 24 hours in advance if and when it wants to discharge waste and iii) ports need to develop a cost recovery system based on fees. The main difference between these requirements and those of MARPOL is that the directive made the development of port waste management plans by ports mandatory, while for IMO this is a recommended feature (Interview European Commission, 2007a). The EU directive is therefore a stronger instrument in ensuring port reception facilities (Interviews European Commission, 2007a; Ministry of Transport, Public Works and Water Management, 2005a).

The second issue concerns TBT in anti-fouling paints. The International Convention on the Control of Harmful Anti-fouling Systems on Ships dealing with this issue was adopted within IMO in 2001. The Convention provides for a ban on the application of organotin based anti-fouling systems as of January 2003. As of January 2008 ships are not allowed to have organotin compounds on their hulls or other external parts or they are required to have a coating that forms a barrier for these organotin compounds to leach into the marine environment. The Convention entered into force in September 2008 after the threshold of 25 states with 25% of world tonnage was reached a year earlier.

In 2003, the EU adopted a Regulation on the prohibition of organotin compounds on ships. This Regulation was meant to have effective and harmonized implementation of the Anti-Fouling Convention within the EU. In reality, there is one element in the directive that is somewhat stricter. However, this element only applies to a small portion of ships i.e. only those ships that have repainted their hull with TBT after July 2003 and then decide to register in the EU. Furthermore, it only applies for the period until all ships are required to be TBT-free (i.e. till 1 January 2008).

4.8.3 Technology-forcing standards within IMO

One change in the global sphere of authority that exists since the 1980s should be mentioned here as well. This is the adoption of technology-forcing standards, instead of technology-following standards. Until the end of the 1990s, IMO conventions and amendments to these conventions have been characterized by adopting standards based on already existing best practices and technologies developed by the shipping industry (Interviews Lloyd's Register, 2007; Transport & Environment, 2007). The only exception was indeed the monitoring and oil in water separators required by the MARPOL Convention itself in 1973 (Interview International Chamber of Shipping, 2007). However, this was undone by the adoption of the Protocol and the elimination of the Load on Top method in 1978.

This situation is repeating itself somewhat, because for ballast water and TBT in anti-fouling there were no best practices and technologies developed yet (Interviews British Chamber of Shipping, 2007; International Chamber of Shipping, 2007; Lloyd's Register, 2007). Thus the debate within IMO on these issues did not only concentrate on what limits to adopt, but also whether viable and cost-effective alternatives existed. For example, with ballast water, the only existing alternative is mid-ocean ballast water exchange, a practice that is very difficult and not without risk. The International Convention for the Control and Management of Ships' Ballast Water and Sediments adopted in 2004 therefore established a review process of other ballast water management options in the years immediately after the adoption, a process that is currently underway. Despite the lack of alternatives and the subsequent lack of progress during negotiations (De La Fayette, 2001), both Conventions contain provisions that are quit ambitious; a ban for TBT in anti-fouling paints and for ballast water the ballast water exchange standard and the ballast water performance standard.

4.9 The authority of the state during the 2000s

The wavering dynamic brought about by the EU sphere of authority also means that the authority of the state in developing steering mechanisms is less 'constant' than before. First, the authority of especially EU states has changed because of the EU sphere of authority (Section 4.9.1). Yet, also the change to technology-forcing standards within the IMO is related to changing authority of the state and the industry. This will be discussed in Section 4.9.2. Finally, the authority of the state in generating compliance after the EU policy change will b discussed in Section 4.9.3.

4.9.1 Authority in developing steering mechanisms after the EU policy change

One of the main changes for the authority of EU member states in the global environmental governance of shipping is that their authority in developing steering mechanisms no longer depends solely on what happens at IMO. There is now another sphere of authority in which standards for shipping are developed, i.e. the EU. This means that these states have gained more control over setting standards for their ships and for ships entering their waters and ports. After all, it has become a more common practice for the EU member states to develop standards on the EU level when they cannot get what they want at IMO.

Of course, this still means that these states' influence on steering mechanism is limited by the political game taking place within the EU. Within the European Council, the same political struggle between maritime interests and environmental interests takes place as within IMO. That is because several of the largest flag states are EU member states, notably Greece, Malta and Cyprus, while on the other hand the North Sea and Baltic states are much more concerned about protecting their marine environment against pollution from shipping (Interview European Commission, 2007b). However, the difference with IMO is that the EU decision making is also influenced by the involvement and positions of the European Commission and Parliament (Pallis, 2006). And until now, these two institutions have advocated the adoption of stricter environmental standards.

In other words, overall the EU seems to be more willing to adopt stringent environmental standards. Until now, EU legislation has therefore accommodated the environmental concerns of port and coastal states better than the IMO conventions. Whether this also leads to an increase in the authority in developing steering mechanisms within the global sphere of authority, depends on the decision making process within IMO.

In those matters in which the initiative of developing EU legislation is taken or the willingness to do so is expressed, the politics within IMO changes. In particular, because (the threat of) EU legislation implicitly entails more stringent environmental standards for EU registered ships and ships visiting the EU. This puts large flag states and the shipping industry, which are not immediately inclined to adopt environmental standards, under pressure to adopt those environmental standards anyhow. Otherwise, they will face a situation in which the level playing field is distorted. The potential of EU legislation is therefore a power resource for the EU member states in the development of steering mechanisms within the IMO. That this is a powerful resource is already demonstrated in the double hull case. Similarly, according to IMO (Interview International Maritime Organization, 2007), in the case of air emissions, the industry has also been keen to satisfy EU's wishes, to stave of unilateral steering mechanisms. According to the International Chamber of Shipping (Interview International Chamber of Shipping, 2007), threats of the EU can even be considered to have a certain element of blackmail in them, especially because it is unclear what the EU wants exactly. This power resource is further supported by the challenging of the existing discourse and the EU acting as a coalition in the negotiations within IMO. The result of more stringent global standards therefore expresses the increase in authority that EU member states have gained during the development of IMO steering mechanisms.

However, the authority of the EU member states during the development of steering mechanisms in the global sphere of authority is a bit more complicated than this. First, the EU coalition not only enhances the influence of EU member states, but at the same time limits the freedom and position of *individual* member states within IMO (Interviews Ministry of Transport, Public Works and Water Management, 2005a; 2008). They no longer have an individual voice, instead, they have to voice a position that is coordinated at the EU level. Thus the authority of individual member states on the development of steering mechanisms is defined by one extra governance layer (Interview Ministry of Transport, Public Works and Water Management, 2008). They have to negotiate within the EU on the position that they together will voice within IMO and then take this position as the starting point for negotiations within IMO. So while acting as a coalition can on the one hand make the environmental position stronger, an individual member state will have to compromise on that environmental position at the same time.

In addition, it is not always seen as legitimate by both the wider shipping community as by European member states to use this alternative of developing steering mechanism for shipping in the EU. In fact, the activities of the EU around the double hull issue and air pollution were often seen as controversial by non-European states and interest groups. For example, according to an UK civil servant (Interview Department for Transport, 2007), the wider shipping community had difficulties with understanding what the EU was, why it emerged in the IMO arena and what the agenda is of the EU in shipping matters. As a result, the EU and its individual member states might lose credibility as a constructive partner in developing sensible standards for shipping. Something that for example the Netherlands has constantly been working on, because within IMO one has to prove oneself before you will be heard and taken seriously (Interviews International Chamber of Shipping, 2007;

Ministry of Transport, Public Works and Water Management, 2008). The increased European activities in governing shipping can therefore have certain drawbacks and repercussions for the position of EU member states within the global sphere of authority.

4.9.2 Authority in developing technology-forcing steering mechanisms

This observed change from technology-following to technology-forcing steering mechanisms is in fact a consequence of the previous policy change through which the role and authority of port states was institutionalized within the global sphere of authority. Both in the issue of TBT and of ballast water, the power of port states to set standards for ships entering their ports has been a driving force behind the adoption of technology-forcing standards.

In the global sphere of authority that existed between the 1950s till the 1990s, the shipping industry often delivered best practices and technologies on which new steering mechanisms could be based. This was even one of the main ways through which the industry exercised authority on the adopted steering mechanisms. They have, however, lost this authority in recent years, when technology-forcing standards were adopted.

In the TBT and anti-fouling case, there were some alternatives such as copper or silicone based anti-fouling systems. However, these alternatives were much more expensive than TBT and less effective. A debate was therefore going on about whether these paints were viable alternatives or not. The paint manufacturing industry, which was represented by the European Council of Chemical Manufacturers' Federation at IMO, supported a global ban of TBT paints (Tan, 2006). They would benefit from an increase in demand for other types of anti-fouling paints which they believed already existed.

In contrast, the shipping industry denied the existence of viable alternatives till the spring of 2000 (De La Fayette, 2001). The shipping industry had grave reservations against the ban, because the existing alternative paints were costly and ships would have to be repainted more often because of paints being less effective in keeping organisms from detaching themselves to the ships' hulls (Tan, 2006). However, the power of the shipping industry was undermined by the willingness of certain states, i.e. the North Sea states and Japan, to impose unilateral or regional regulations. As this threatens the level playing field in shipping, the shipping industry was compelled to support the global ban, to stave off the less evil option of unilateral and regional regulations (Tan, 2006). The lack of consensus over viable alternatives stood therefore not in the way of adopting a global ban on TBT or other organotin substances in anti-fouling.

Another way in which the environmental coalition of Japan, the North Sea states and the environmental NGOs put their mark on the Anti-Fouling Convention was by lowering the entry into force requirements to 25 states with 25% of the world tonnage instead of the tradition 50% of the world tonnage. This decreased the dependence on the group of largest flag states to ratify the Convention (Tan, 2006).

The ballast water management issue was initiated and led by Australia, the US and Norway. Especially Australia is vulnerable to alien species because of its geographical uniqueness and vulnerability of its flora and fauna (Tan, 2006). An important element in the debate on eliminating alien species is the management options available. Ballast water exchange in the open sea is one of these options, but is opposed by the shipping industry and large flag states because of safety concerns and doubts about its costs and delays (De La Fayette, 2001; Tan, 2006). There are other options that are used on land, but it is a question mark whether these can be used on ships and how effective they are. That is why the shipping industry and flag states are proponents of a system where ballast water management systems have to go through an approval system of IMO (De La Fayette, 2001).

Again, the power of the flag states and shipping industry was countered by the already existing national, sometimes voluntary guidelines and regulations. The maritime actors were faced with the dilemma between supporting a global steering mechanism or having to deal with differences in standards across countries (De La Fayette, 2001).

In the development of technology-forcing standards, flag states and the industry have lost authority, because they are overruled by port states that develop unilateral standards. In both the TBT issue and the ballast water issue, the shipping industry was under pressure because national and regional steering mechanisms already existed or were to be developed. This shows the authority of those port states that were experiencing the adverse effects from TBT or alien species over the authority of the industry.

Yet, while the Anti-Fouling Convention entered into force in September 2008 and compliance with the Convention is possible because alternatives exist, the Ballast Water Management Convention seems to suffer from its technology-forcing provisions (Interviews European Commission, 2007b; Friends of the Earth International, 2007; International Chamber of Shipping, 2007). The approval of ballast management systems is an ongoing process within IMO, where several management systems have received basic or final approval, but none have received the necessary type approval yet (Marine Environment Protection Committee, 2008). In addition, the earliest implementation date mentioned in the Convention is 2009 and Norway together with the shipping industry has been proposing to exempt ships from compliance when management options are not available at that date (Interview Friends of the Earth International, 2007; Marine Environment Protection Committee, 2006). Subsequently, it was decided to adopt a resolution calling on states not to apply the Convention for a limited period of time (Marine Environment Protection Committee, 2007). Thus even though the Convention is adopted, considerable work still needs to be done before it is possible to implement and enforce the Convention.

The adoption of these two steering mechanisms was about finding a delicate balance in setting ambitious standards to stimulate technological developments and best practices on the one hand and relying on existing technologies and best practices to set standards. While in the past, the IMO chose to take the latter route, the TBT and ballast water cases show that IMO is becoming more proactive. The choice of a goal-based standard in the revision of MARPOL Annex VI on air emission is also part of this trend. The rationale behind the goal-based standard is that the standards are set, but that the way in which these standards will be achieved is open to the states and shipping industry (Interview Ministry of Transport, Public Works and Water Management, 2008). This stimulates states and the industry to be innovative.

4.9.3 Authority in generating compliance

The influence of the EU on the global sphere of authority is not limited to the development of steering mechanisms only. Its influence is also noticeable in the compliance with steering mechanisms within the global environmental governance of shipping. The EU issued a Directive on Port State Control, which directly affected the Paris MoU and the way Port State Control is conducted by the EU member states.

The mandatory feature of the Port State Control Directive entails that EU member states have to transpose the directive in national law. Implementation of the EU directive is also subject to control of the European Commission. When member states fail to implement the directive properly, the EC can start the infringement procedure which can end with the imposition of a considerable fine by the European Court of Justice. It should, however, be noted that this infringement procedure is a compliance mechanism in the European sphere of authority and not the global sphere of authority.

Still, despite the infringement procedure, the EU member states have not always implemented the directive and thus the Paris MoU properly. For example, it took till 2004 for all members of the EU and the Paris MoU to reach the 25% inspection rate (Paris Memorandum of Understanding on Port State Control, 2005). In addition, as Frank (2005) observes, only 5 of the 15 EU members had transposed the amendments of 2002 in their national law before the deadline of 22 July 2003. And that this is still an issue is exemplified by the fact that the EC has indeed started the infringement procedure against Malta and Italy in 2005 and against Portugal in 2006 (European Commission, 2005, 2006).

The main change for the authority of the state in generating compliance is not so much the fact that the European states started to undertake and coordinate their port state control efforts; they have already done so since 1982. The difference is rather that the EU member states are now subject to a mandatory steering mechanism that compels them to arrange a port state control system instead of a voluntary MoU. The effect on the wider global sphere of authority is that over the years, a consistent system of port state control has emerged in Europe affecting foreign ships trading with the EU. This has therefore increased the authority of (EU) port states in the global environmental governance of shipping.

It was already observed that the EU also has a record of developing directives and regulations to ensure harmonized implementation of IMO Conventions by EU member states. Examples are the Directive on Port Reception Facilities, the Directive on Organotin compounds on ships and the Sulphur Content in Oil Directive (which implements Annex VI of the MARPOL Convention). Similar to the Directive on Port State Control, it is not so much the implementation by individual member states per se that is affected by this directive, but the fact that these directives are mandatory and member states are subject to control by the EC is the main change. Moreover, the manpower of the EU to control compliance has increased with the establishment of the European Maritime Safety Agency (Interview European Commission, 2007a). In other words, while the ratification and implementation of international conventions is still more or less voluntary, the EU directives make them compulsory and give the EC the means to check whether the member states comply. What not changes is the fact that individual (member) states are responsible for implementing the international en European standards.

Surprisingly, some EU member states have not ratified the international conventions even though they are required to implementation the standards of these conventions through the European directives. For example, the Anti-Fouling Convention is not ratified by several EU member states (i.e. Belgium, Germany, UK, Portugal) while those states are already required to implement the provisions of the convention through the Directive on organotin compounds on ships since 2003. According to a Dutch governmental official (Interview Ministry of Transport, Public Works and Water Management, 2008), this might be related to the fact that ratification is sometimes subject to delays because of the need for parliamentary consent or because of technicalities with regard to the transposition into national law.

Since most of these directives pose additional requirements for ships entering European ports, the effect of these directives is not only felt in Europe, but also by foreign ships trading with EU member states. These directives are therefore often more than a compliance mechanism in the European sphere of authority. Through the implementation of these directives and the link with port state control, ships entering a European port are controlled on compliance with issue covered by these directives, e.g. TBT and anti-fouling.

All in all, EU directives have increased the authority of the EU port states over those ships entering EU ports. In other words, EU port states are important in generating compliance within the global sphere of authority. Moreover, the EC itself has gained authority over compliance, through the infringement procedure. Yet, this authority is over EU member states and within the EU sphere of authority rather than directly onto ships.

Besides the EU, the IMO itself has been working on enhancing implementation and compliance as well. For this purpose, the Flag State Implementation Sub-Committee was established in 1992. One of the compliance mechanisms they have initiated was the self-assessment of flag state performance form, which was adopted in 1999. However, this self-assessment form did not get the expected response, i.e. in 2001 only 32 states had completed the form (International Maritime Organization, 2001).

A follow-up on the self-assessment form was developed in the form of a voluntary audit scheme, adopted within IMO in 2005. Through this scheme a party to IMO can participate on a voluntary basis in an audit in which the implementation and enforcement of IMO regulation in that state is assessed. The voluntary nature of the audit scheme means that it is only done at the initiative of a member state. Seeing the experiences with the self-assessment form, the question is to what extent states will be taking the effort to do the audit. Just like with the self-assessment form, the power behind the scheme is limited because it relies on the state itself to initiate the audit. However, once an audit is done, the outcome will be able to help the member state to better implement IMO regulations.

Reporting of enforcement efforts for the MARPOL Convention remains an issue, which is discussed by the Flag State Implementation Committee as well. For example, in 1999, only 25 states submitted reports to IMO and in 2001 only 29 out of 162 states did (International Maritime Organization, 2001, 2003). However, urging states to report is all that IMO and the Flag State Implementation Committee can do.

Recently, the focus of the Sub-Committee on Flag State Implementation is also on the inadequacy of port reception facilities. This inadequacy is seen as one of the main reasons for non-compliance of the MARPOL Convention. An action plan was adopted by the MEPC in 2006 and standardizing of among others the notification and delivery forms are currently aimed for. Another element of the Sub-Committees work on this issue is the development of the IMO Global Integrated Shipping Information System Port Reception Facilities Database to provide information and contact points for states on available reception facilities.

Despite these initiatives by the IMO, the authority of IMO in generating compliance has hardly increased. The difficulty is that IMO depends on its parties to develop compliance mechanisms that are mandatory and to which parties can be hold accountable. The main mechanisms in ensuring compliance that IMO therefore has is exercising political pressure (Interview Ministry of Transport, Public Works and Water Management, 2005a). As long as this interdependence exists, the compliance mechanisms of IMO will remain functioning on a voluntary basis and will not increase the authority of the IMO over the compliance of its parties.

4.10 Future outlook: initiatives outside IMO and EU

Within the environmental governance of shipping, new initiatives worth mentioning are taking place outside IMO and the EU. These are initiatives of the industry, the environmental NGOs, classification societies, ports and even cargo owners. However, it should be noted that these initiatives are most of the times in their early stages and in terms of the analysis done in this chapter, not worthy of the label sphere of authority or policy change. Of course, the question remains whether they will become a policy change in the future.

The industry as a whole, represented by the ICS, can still be regarded as conservative (Interview European Commission, 2007b) or as taking a compromised position at best (Interviews British Chamber of Shipping, 2007; Department for Transport, 2007). This has to do with the fact that the ICS has a large membership of many parts of the worlds and has to find a common position in representing its members. Moreover, the part of the industry that owns or operates older ships, simply is not able to be at the forefront of the industry in terms of environmental performance (Interview Department for Transport, 2007). Yet, on a national or individual level, some industry actors are much more proactive (Interviews Department for Transport, 2007;

European Commission, 2007b). For example, Wallenius Wilhelmsen invests in developing and testing new technologies and is using low sulphur fuel across its entire fleet (Interview Department for Transport, 2007). BP and Stolt-Nielsen Transportation Group are looking into waste separation systems onboard (Interviews Department for Transport, 2007; Stolt-Nielsen Transportation Group, 2007). BP and P&O are also trying scrubbers to clean their emissions from sulphur (Interview Department for Transport, 2007). Maersk already reduced their NOx emissions to below required levels (Interview European Commission, 2007b). Moreover, companies are implementing environmental management systems to integrate the attention for environmental issues throughout the whole company (Interviews British Chamber of Shipping, 2007; Department for Transport, 2007; Stolt-Nielsen Transportation Group, 2007).

Of course, some of these initiatives can be regarded as an old practice, because companies have always tried to develop new technologies to make it easier to comply with international standards. However, current initiatives go further than that and seem to show that the attitude of companies to environmental issues is changing. They have become more proactive and willing to show that they are working on solutions for environmental pollution (Interview British Chamber of Shipping, 2007). Still, the most convincing example is that Intertanko has taken an environmental position in the recent debate on the revision of MARPOL Annex VI on air pollution. As IMO explains (Interview International Maritime Organization, 2007), Intertanko's call to completely switch to distillate fuels, which contain much less sulphur, is a radical change from its previous position with regard to air pollution.

One of the driving forces behind this change is the realization that the environmentally friendly image of shipping is under threat (Interview Friends of the Earth International, 2007). For example, the upsurge in awareness about climate change since Al Gore is also affecting the shipping industry (Interview British Chamber of Shipping, 2007). The shipping community feels the pressure to deal with CO_2 emissions from shipping, because otherwise the CO_2 emissions from shipping will surpass those of other types of transportation. Moreover, now that the public is focusing on climate change the reputation and image that shipping currently has is under threat. This is something that the industry is slowly becoming aware off. The industry and others in the shipping world are realizing that they have to take efforts to ensure that shipping stays the most environmental friendly mode of transportation.

Lloyd's Register and other classification societies are supporting this trend. Ship owners are asking Lloyds to help them to go beyond regulations (Interview Lloyd's Register, 2007). At the same time, Lloyd's Register has published its own environmental standard, the Environmental Protection Notation, in 1998. This environmental standard invites ship owners to go beyond the international requirements.

The environmental NGOs the North Sea Foundation has developed the Clean Ship Concept to trigger debates about the future of shipping. The start of this concept was marked with a symposium organized by the North Sea Foundation in 2002 under the title 'Zero-emission ship: utopia or seaworthy?' where 70 maritime specialists met and agreed that a Clean Ship could be possible before 2015 (North Sea Foundation, no date). After that, the Clean Ship Concept was put forward by the North Sea Foundation during the 2002 North Sea Ministerial Conference and the Ministers agreed to include the concept in their declaration. Since 2006, there is a website specifically devoted to the Clean Ship Concept. The website is supported by both the North Sea

Foundation and the Dutch Ministry of Transport, Public Works and Water Management. In addition, North Sea Foundation has informal agreements with the International Chamber of Shipping to elaborate the concept in the updated version of the Environmental Code of Practice of ICS (Interview International Chamber of Shipping, 2007).

Another initiative is closely related to ports. The Green Award Foundation has been initiated by Rotterdam Port and the Dutch Ministry of Transport, Public Works and Water Management, but currently has a committee with members from industry associations, ports associations, an environmental NGO and a classification society. A Green Award is awarded to individual ships that meet the Green Award requirements. Currently, over 200 ships (of about 38 ship-owners) carry the Award (Greenaward, 2009b). Moreover, ports in some countries (the Netherlands, Belgium, Lithuania, Spain, Portugal, South Africa and in New Zealand) have started to give a differentiated port fee to ships that carry a Green Award (Greenaward, 2009a).

Finally, a new trend is that cargo-owners require that those ships that transport their goods meet certain requirements. Especially companies that produce goods with a somewhat dirty reputation, such as oil, cars and nuclear waste, tend to set requirements for the ships that transport their goods (Interviews International Maritime Organization, 2007; Lloyd's Register, 2007). The driving force behind this is the environmental management systems and the chain management that those cargo-owners implement (Interview Department for Transport, 2007). These cargo-companies do not only look at their own company, but also at the companies that are transporting their goods, i.e. the ship owners. Similarly, the Oil Companies' International Marine Forum has developed a Tanker Management Self-Assessment Program, which includes environmental requirements, as a tool to assess the safety and environmental credentials of tankers. This can help the oil companies to choose ships that meet a certain level of safety and environmental performance (Interview Stolt-Nielsen Transportation Group, 2007).

All in all, these initiatives show that different industry stakeholders (individual, but large ship-owners, cargo owners) and non-industry stakeholders (environmental NGOs, classification societies and ports) have each taken on the challenge to become more proactive and more innovative in combating environmental pollution from shipping. These initiatives are taking place outside the established spheres of authority, but are still very fragmented. It remains to be seen whether a more coherent, private based sphere of authority will emerge within the environmental governance of shipping in the future, or whether these will remain initiatives that support, but not challenge or compete with the IMO sphere of authority. If such a private-based sphere of authority would emerge, it could radically change the political landscape in the environmental governance of shipping.

Chapter 5.
Changing authority in the environmental governance of offshore oil and gas production

5.1 Oil and gas production on the North Sea

In July 1959, a huge gas field was discovered in the north of the Netherlands. This and unease with the growing dependence on Persian Gulf oil sparked the desire to look for oil (and gas) in the shallow parts of the North Sea. The quest for offshore oil and gas fields started in the early 1960s with the first seismic researches. Some years later, in September 1965, the first gas field was struck on the UK continental Shelf and in October 1969 on the Dutch Continental Shelf. In the summer of 1970, the first oil field was discovered on the Norwegian Continental Shelf and in 1971 on the British Continental Shelf. These discoveries marked the beginning of oil and gas production in Europe and especially in the Netherlands, Norway and the UK.

Since these offshore activities were new and taking place outside territorial borders on the North Sea, there was no clear regulatory framework for allowing companies to drill, search for and produce oil or gas. Fortunately, questions over what the boundaries are of nation's continental shelf, who owned the seabed of the continental shelf and who would be allowed to exploit the seabed were answered in 1964 when the Convention on the Continental Shelf entered into force (Nelsen, 1991). This Convention was the predecessor of the United Nations Convention on the Law of the Sea which was adopted in 1982. This Convention allowed coastal states to extent sovereignty to the continental shelf and to exploit the resources in that area.

One issue for the governments of the Netherlands, Norway and the UK remained however. They still had to set the conditions under which exploration and production of oil and gas could take place (Nelsen, 1991). All three states adopted Mining Laws to set these conditions: the UK amended the Petroleum Production Act in 1964 to cover offshore gas and oil activities; the Netherlands adopted the Mining Law for the Continental Shelf in 1965; and Norway adopted a Royal Decree in 1965. These regulations require oil companies to have a license for exploration and for production. Only the Netherlands also requires a license for seismic surveys.

Besides having these laws and licenses for gas and oil exploration and production, a desire to set environmental conditions for the production of oil and gas started to develop in the 1970s. This chapter will analyse which spheres of authority set and implement these conditions and how these spheres of authority have changed over the years.

5.2 The emerging regional sphere of authority during the 1970s and 1980s

Even though a surge of environmental awareness arose in the early 1970s, environmental concerns received little attention in the offshore exploration of oil and gas (De Jong, Weeda, Westerwoudt & Correljé, 2005). Governments were much more concerned with how to deal with this new industrial activity than with environmental effects of oil and gas exploration. The consequence is that a sphere of authority governing environmental pollution from oil and gas production in the North Sea only developed slowly. The first steering mechanism was not adopted before 1978. This steering mechanism was adopted within the regional Paris Convention on the Prevention of Marine Pollution from Land-based Sources. This means that the first sphere of authority that emerged in the environmental governance of oil and gas production offshore was a regional one.

5.2.1 The Paris Convention and oil pollution

The development of the regional sphere of authority that governed the environmental effects from offshore oil and gas production started at the end of the 1970s. At the heart of this sphere of authority was the 1974 Paris Convention on the Prevention of Marine Pollution from Land-based Sources. Even though this Convention focused on onshore sources of pollution, the offshore industry was covered by this Convention as well. This Convention is a regional Convention covering the North-east Atlantic and signed by Belgium, Denmark, France, Germany, Iceland, Luxembourg, the Netherlands, Norway, Portugal, Spain, Sweden, Switzerland and the United Kingdom. The European Commission was allowed to become a Party as well. The Paris Convention entered into force in 1978.

The Paris Convention set down rules of the game for the development of steering mechanisms. The Paris Commission adopted steering mechanisms in the form of binding decisions, or non-binding recommendations and agreements. The Paris Commission met annually and was supported by two working groups: the joint monitoring group and the technical working group. The latter had a number of subgroups, one of which was the Working Group on Oil Pollution (GOP). The Paris Commission decided that the technical working group and its subgroups would limit their activities to technical and scientific matters, while the Commission itself would take the political decisions. Representatives of national delegations to the working groups would therefore preferably be scientists and not necessarily representing national interests (Environmental & Safety Consultancy, 1992).

Even though it was agreed that the working groups would focus on scientific and technical issues, in practice, the GOP prepared the decisions, recommendations and agreements which would in turn be adopted by the Paris Commission (see GOP summary records 1984-1995). This means that not only debates on the extent of environmental impacts but also negotiations and building consensus for future steering mechanisms took place in the working group.

The actors that were active in the GOP were first and foremost the Parties to the convention, i.e. the states. Observers such as Finland and representatives from other regional and international conventions (e.g. the Baltic Marine Environment Protection Commission or the International Council for the Exploration of the Sea) were present as well. Industry associations or other societal actors were not allowed to become observer. This does not mean that they were not part of this regional sphere of authority. The oil companies that are active in oil and gas production offshore, i.e. the offshore industry, was at that time represented by the Oil Industry International Exploration and Production Forum (E & P Forum). The E & P Forum has been actively contributing to decision making in the Paris Convention in several ways. First, industry associations and other NGOs were allowed to give presentations before the meeting of a GOP commences, an option that was also used by the E & P Forum. Second, special meetings were held with the E & P Forum to discuss specific issues. Environmental NGOs were not active at the GOP meetings in the 1980s and early 1990s (see GOP Summary Records 1984-1995).

Yet, the Paris Commission and its working groups was not the only forum where steering mechanisms for the environmental governance of the offshore industry were discussed. The North Sea Ministerial Conferences of 1984 and 1987 were important in the development of steering mechanisms as well. As explained in the previous chapter, the NSMCs are a forum at which the North Sea states discuss issues that threaten the marine environment of the North Sea. More importantly, the outcome of these conferences is a political declaration in which existing international forums are supported and urged to take action. These political declarations are not considered to be a steering mechanism, but did lead to the development of steering mechanisms by other forums. For the offshore industry, the Paris Commission was the international forum that was asked to adopt steering mechanisms.

In the steering mechanism dimension, several mechanisms have been adopted since 1978. Immediately at the first meeting of the Paris Commission in 1978, a steering mechanism was adopted focusing on oil pollution by the offshore industry. The offshore industry discharges oil in production, drainage and displacement water. Production water is water that is extracted during gas and oil production together with the oil or gas itself and is the main source of oil pollution. The Parties to the Paris Convention agreed to a provisional target standard of 40 parts of oil per million parts of water for any discharge (usually referred to as 40 ppm). This 40 ppm was an average and not an absolute standard. In 1984, when the first North Sea Ministerial Conference was held in Bremen, the political declaration also mentioned produced water as an issue. The NSMC asked for regular data that covers the amount of oil spilt into the North Sea and the reasons why (North Sea Ministerial Conference, 1984). It also agreed on a guiding value of 40 ppm hydrocarbons in the discharge of produced water. The response of the Paris Commission came in 1986, when it was agreed that this provisional standard became a target standard for both existing and new platforms. Moreover, in 1988 a requirement for platforms to report when they are unable to meet the standard was added.

Another element for minimizing oil pollution did not come from the Paris Convention, but from Annex I of the MARPOL Convention which plays an important role in the environmental governance of shipping. This Annex also set discharge limits for fixed and floating rigs. This meant from the moment that the MARPOL Convention entered into force in 1983, fixed and floating rigs were not allowed to discharge oil or oily mixtures with more than 100 ppm of oily content. In territorial waters and in special areas the limit was set on 15 ppm of oil in mixtures.

Nevertheless, by the time that the MARPOL Convention entered into force, the Paris Commission already adopted a provisional target of 40 ppm of oil in drainage, production and displacement water. Moreover, the E & P Forum wanted to exclude fixed platforms from MARPOL and therefore submitted a paper to the Marine Environment Protection Committee of the IMO in 1984 (Marine Environment Protection Committee, 1984). It was subsequently decided that only machinery space drainage discharges would be subject to the MARPOL Convention.

The issue reducing the oil content of produced water came back on the agenda again in 1990 during the third NSMC in The Hague. The Paris Commission was asked to indicate whether a 30 mg/l standard would be technologically feasible. In order to do so, an ad hoc working group to consider the oil content in produced water was organized in 1991. During this ad hoc working group, existing technologies were reviewed. The conclusion was that the 30 ppm target was not technological feasible yet and that the 40 ppm remained the appropriate target (Secretariat of the Paris Commission, 1991).

In the early 1980s, another source of oil pollution became an issue, namely the use and discharge of oil based muds. Muds are used as drilling fluids. Drilling fluids have several functions during drilling: they lubricate and cool down the drill bit, they transport the small pieces of rock that break of during drilling (drill cuttings) back to the platform and they balance the pressure of fluids in the rock formation to prevent blowouts. Drilling muds which could not be reused and cuttings contaminated with oil were discharged into the sea (Gray, Bakke, Beck & Nilssen, 1999).

However, during the early 1980s realization grew that oil based muds and the oil contaminated drill cuttings were much more polluting than originally assumed. In the course of the 1980s a number of steering mechanisms were developed to reduce the discharge of oil based muds and cuttings. For example, the NSMC of 1984 in Bremen agreed that the discharges of drilling muds and cuttings should be limited and avoided, especially of oil based muds (North Sea Ministerial Conference, 1984). This ran parallel with the first restrictions on the use and discharge of oil based muds and oily cuttings that were arranged in a decision of the Paris Commission in 1984. This decision entailed a ban on the discharge of oily muds and a reduction target for residual oil on cuttings. Two years later, the Paris Commission approved a decision to restrict the use of diesel oil muds to certain exceptional circumstances.

During the NSMC of 1987 in London, the limitation of the use of oil based muds was again mentioned in the political declaration. It, for example, asked the Paris Commission to establish regulation to reduce the oil content of discharged cuttings (North Sea Ministerial Conference, 1987). The response from the Paris Commission came in 1988 with the adoption of a decision to arrange the request of the NSMC. This decision prohibited the use of oil based muds for the upper part of the well and it limited the discharge of oily cuttings to cuttings with an oil content of 100 g/ kg dry cuttings.

The third NSMC in 1990 again discussed the issue of muds and cuttings. The Paris Commission was asked to coordinate the development of national action plans to prohibit the discharge of oil contaminated cuttings and to agree on a date when the discharge of oil contaminated cuttings would become prohibited (North Sea Ministerial Conference, 1990). The Paris Commission indeed restricted the discharge of oily cuttings further in 1992 when the decision was adopted that allows not more than 1% oil on cuttings. In practice, this decision meant a ban on the discharge of oily cuttings; they have to be transported to land instead.

However, the steering mechanisms for the issue of muds and cuttings were not only discussed in the NSMCs and the GOP but also in several meetings between the GOP and the E & P Forum. For example, in 1983, the industry explained that there was a need for using oil based muds in some cases and how they reviewed the environmental effects of discharged oil based muds (Secretariat of the Paris Commission, 1984). The representatives of the GOP also presented their findings about these effects. In 1985, such a meeting was organized again on the initiative of the Paris Commission. This time, the environmental effects of and treatment systems for oily cuttings were discussed as well (Secretariat of the Paris Commission, 1986). Such meetings usually ended with agreed facts, which were then used in the GOP meeting to discuss the issue further. In 1988, the developments since 1985 and the agreed facts were reviewed.

Finally, another steering mechanism that was related to the use of drilling muds was the 1981 decision on the notification of chemicals used offshore. While this decision did not set any restriction to the use of chemicals, it did require Parties to the Convention to be informed of the scale of use, composition, biodegradability and toxicity of chemicals used in drilling muds. It was not until 1987 that the gathering of this information was raised to the regional Paris-level: a questionnaire was designed by the GOP to gather information about chemicals used offshore (Secretariat of the Paris Commission, 1987). A report on the discharges of chemicals from offshore installations has been developed annually since 1990 (Secretariat of the Paris Commission, 1990).

Reducing discharges of hazardous chemicals was addressed for the first time in the second NSMC in 1987. The North Sea states asked the Paris Commission to prohibit or strictly limit the discharge of chemicals that have a potential risk for the marine environment (North Sea Ministerial Conference, 1987). In 1990, the third NSMC requested the development of a harmonized mandatory control system for the discharge and use of offshore chemicals under the Paris Convention (North Sea Ministerial Conference, 1990). In 1994, a decision was adopted containing two lists of substances and chemicals. The first list of substances would be subject to strong regulatory control, while the discharge of substances of the second list would be prohibited. This decision was superseded by the 1996 decision through which the Harmonized Mandatory Control System for the Use and Reduction of the Discharge of Offshore Chemicals was adopted.

It is of course no surprise that issues discussed within the working group on oil pollution overlapped with issues discussed during the NSMC. After all, the North Sea states are actors in both forums. The differences between these NSMCs and the Paris Convention are the scope of the area (North Sea versus North-East Atlantic), the number of members, the scope of the issues discussed and the outcome.

While the NSMCs have a holistic approach towards the environmental protection of the marine environment (De La Fayette, 1999), the Paris Convention focuses only on land-based sources. The outcome also differs, because the NSMCs come with political declarations that call for certain measures by existing international institutions. Together with the higher political profile of the NSMC, the political declaration in fact gives a political impetus to the development of steering mechanisms under, among other, the Paris Convention (De La Fayette, 1999). Thus as Tromp and Wieriks (1994, p. 622) explain 'due to this close relationship, agreement by the Ministers about the North Sea area could be followed relatively quickly by agreement in the Commissions for the North East Atlantic Area'.

Skjearseth (2006) argues that the (partly) overlapping participation in both bodies is important to explain the dynamics between the Paris Convention and the NSMCs. He argues that the exclusion of the Mediterranean states within the NSMCs left only the UK as a laggard among the North Sea states. It was therefore easier to put pressure onto the UK and to build consensus over certain issues and measures during the NSMCs.

The North Sea Ministerial Conferences therefore play a role in the power dimension of the regional sphere of authority. These conferences allowed the North Sea states to discuss those issues that were especially relevant for the North Sea and to build consensus on the way forward in these issues. This has more than once led to a speedy adoption of Paris decisions. It could of course be argued that the most important states in the GOP were already North Sea states and consequently the NSMCs are not more than an extension of the GOP. However, the NSMCs were with fewer participants and with a focus on the North Sea, which have made it easier for Norway, the UK or the Netherlands Sea states to argue their case. On top of that, the fact that the NSMC is a forum at which Ministers express their will to have certain regulations adopted or developed gives the work in the GOP and the Paris Commission an important political impetus (Hey, 1993).

As noted before, industry associations were allowed to give presentations at the beginning of GOP meetings. Thus the negotiations between states on steering mechanisms for oil pollution were preceded by input from the industry. Based on the Summary Records of the GOP meetings between 1984 and 1995 it can be concluded that initially, it was only the E & P Forum who occasionally did that. In the early 1990s, not only the E & P Forum, but also the European Oilfield Speciality Chemicals Association (EOSCA) became a regular presenter at the GOP meetings. EOSCA was established in 1990 and represents the chemical manufacturers and chemical service companies who are involved in the chemicals used offshore. During the years in which the GOP was deliberating a harmonized Mandatory Control System for Chemicals, EOSCA and the E & P Forum held joint presentation at every GOP meeting.

It should also be noted that in the regional sphere of authority, the only stakeholders for which the governments were susceptible were the oil and gas producing companies and not environmental NGOs. That is because the national delegations were not lobbied by environmental NGOs (Department of Trade and Industry, 2006c; Hydro, 2006). This means that environmental NGOs were not part of the actor dimension of the regional sphere of authority and that there was no competition between the views of the industry and those of the environmental NGOs.

In 1992, the Paris Convention was superseded by a new Convention, i.e. the Convention for the Protection of the Marine Environment of the North-East Atlantic. This Convention is known as the OSPAR Convention, because it replaces both the Oslo and Paris Conventions. The Oslo Convention for the Prevention of Marine Pollution by Dumping from Ships and Aircraft was adopted in 1972. The Commissions of the 1972 Oslo Convention and the 1974 Paris Convention had been cooperating with each other since their establishment. The OSPAR Convention entered into force in 1998.

According to Hey (1993), there were several reasons why the decision to merge and update the Oslo and Paris Conventions was taken. First, because of developments in marine environmental policy, the work of both Commissions no longer corresponded to the Conventions. For example, the traditional black/grey list approach of the Oslo Convention was no longer compatible with the sector approach that had emerged. Second, the Commissions sometimes had to duplicate their work on procedural and financial matters, while increased coordination was necessary on other issues.

The new OSPAR convention that superseded the Paris Convention also means a change of the institutional basis in the regional sphere of authority that governs the environmental effects of offshore oil and gas production. One of the main changes is the broadened framework – the new Convention covers all sources of pollution – in which the steering mechanisms for the offshore industry would get adopted in the future. However, the rules of the game that guide the decision making process or the content of the steering mechanisms adopted did not change. The OSPAR Commission still has the duty to adopt decisions, recommendations and agreements to prevent pollution of the marine environment, including from the offshore industry. This Commission also supervises the implementation and effectiveness of the adopted steering mechanisms.

Even though the number of working groups increased, there is still a single working group that does the preparatory work and advises the OSPAR Commission on pollution from the offshore industry. In 1995, the name changed from the Working Group on Oil Pollution to the Working Group on Sea-Based Activities (SEBA). This change in name was related to the new OSPAR Convention. The new working group was a combination of the Working Group on Oil Pollution under the Paris Convention and the Standing Advisory Committee for Scientific Advice which advised both the Paris and Oslo Commissions (Secretariat of the Oslo and Paris Commissions, 1995). This meant that the working group would not only cover pollution from the offshore industry, but also discussed sea-based dumping practices which were previously covered by the Oslo Convention.

One of the main changes arising out of the OSPAR Convention is a change in the rules of the game which has an impact on the actor dimension. The OSPAR Convention allows for NGOs to become observers during the meetings of the Commission. This means that a common practice had now found a solid legal foundation in the Convention (Hey, 1993). However, NGOs could only become observer at meetings of the Commission and not at the Working Groups (OSPAR Secretariat, 2006). Existing practices of NGOs, i.e. the E & P Forum and EOSCA, presenting their views at the beginning of each meeting of a working group would therefore continue, but full participation was not (yet) allowed.

5.2.2 National implementation: the UK, Norway and the Netherlands

Just like in any other international agreement, the Parties to the agreement – the states – are responsible for implementing the agreed steering mechanisms. This section will discuss the way in which the three main states with offshore industry – the UK, Norway and the Netherlands – have implemented the agreed standards through national law and practices. The compliance mechanisms that are developed within this national context will be reviewed as well. The relationship between the regional sphere of authority and national implementation is visualized in Figure 4.

Yet, besides this implementation and compliance through the national actors, there are several regional compliance mechanisms which should be noted first. These compliance mechanisms are part of the agreed decisions, recommendations and agreements. For example, in 1980 it was agreed that the provisional 40 ppm standard should be based on at least 16 samples per months at pre-arranged intervals on platforms that discharge continuously. The performance of platforms against the target could be assessed based on these samples. The number of platforms exceeding the 40 ppm had to be reported to the Paris Secretariat.

In 1987, this compliance mechanism was further harmonized by the adoption of Paris sampling procedures which increased the reliability of the monitoring data (Secretariat of the Paris Commission, 1988). The information generated through this sampling and monitoring was published in an annual 'report on discharges from offshore exploration and exploitation installations' (see Summary Records GOP 1984-1994). This report dealt with the 40 ppm limit, but also with other sources of oil pollution such as via cuttings and accidental spills. This report is subject of discussion during each GOP meeting, making it among others possible for Parties to assess each others level of compliance.

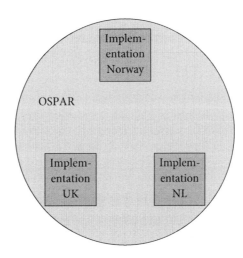

Figure 4. The regional sphere of authority and implementation in Norway, UK and the Netherlands.

On top of that, a triennial report on the discharges from platforms is published, which gives an overview of discharges from platforms over time. This on the one hand allows an assessment of compliance of each Party of the Paris Convention by the Paris Commission. On the other hand, it allows seeing the trends in reductions or increases in discharges. This helps in deciding whether existing measures are effective and whether new measures should be taken.

During the 1970s and 1980s, the Department of Energy was the governmental body that was responsible for the United Kingdom's offshore industry. In 1992, the Department of Trade and Industry (DTI) took over the work of the Department of Energy. However, national rules of the game did not allow these Departments to develop regulations. The only governmental steering mechanisms that could be developed were Acts and other statutory regulations through the Parliamentary decision making route. This meant that the role of Department of Energy/DTI was more an administrative than a regulatory one and that development of new steering mechanisms – to implement the Paris decisions and recommendations – was time consuming.

The offshore industry has been represented by the United Kingdom Offshore Operators Association (UKOOA) since 1973. Although originally established in 1964 as an informal discussion forum for oil companies operating in the UK, by 1973 it had become a permanently staffed organization that represented the offshore industry among others with the government (Nelsen, 1991).

The implementation of Paris decisions and recommendations in the UK was done through a combination between legislation and voluntary agreements with the industry. On the legislative side, the UK adopted the Prevention of Oil Pollution Act in 1971, which made discharge of oil or oily mixtures into the sea illegal. This Act mainly served to regulate oil pollution from shipping, but was considered broad enough to be able to cover offshore oil and gas activities as well (Department of Trade and Industry, 2005; Interview Department of Trade and Industry, 2006b). However, since a total ban on operational oily discharges was untenable given the technology available, exemptions for this ban were made possible under certain conditions in 1975 (Rowan-

Robinson, 2000). Operational discharges from the petroleum industry were indeed exempted from this requirement, while accidental oil discharges were still covered by the Prevention of Oil Pollution Act. Therefore, for the offshore industry, the Act did not immediately serve as a means for reducing oil pollution from operational activities.

The first regulatory effort with regard to minimizing oil discharges from oil and gas production is a statement made by the UK Department of Energy in 1977. This statement implied that platforms are normally required to maintain the oil content of any discharge below an average of 40-50 ppm en below 100 ppm for 96% of the time (Fitzmaurice, 1978). Before taking up this requirement as a condition for the exemption under the Prevention of Oil Pollution Act 1971, the UK proposed to adopt the target of an average of 40 ppm oil in any discharge under the Paris Convention (Read & Blackman, 1980). This limit was, as mentioned above, indeed adopted as a provisional target in 1978. After that, the 40 ppm limit and the 100 ppm limit for 96% of the time became one of the conditions for the exemption under the Prevention of Oil Pollution Act (Environmental & Safety Consultancy, 1992).

A voluntary Offshore Chemical Notification Scheme was adopted in 1979. This voluntary agreement between the government and industry was a way to give guidance to the industry about which chemicals were preferred by the government (United Kingdom, 1986). At the same time, the scheme would lead to more insight in the use and discharge of chemicals both for the government and the individual operators. Again the initiative of the UK was also copied by the Paris Convention. After all, a decision was adopted in 1981 in which the oil and gas industry of the UK, Norway and the Netherlands were asked to notify authorities about the chemicals they used offshore.

According to DTI (Interview 2006a), the voluntary chemical notification scheme was effective because operators did not like to tell DTI when they used a large amount of bad chemicals; instead they preferred to use more environmental friendly chemicals. Thus an incentive for companies to reduce the use of the most hazardous chemicals was built into this steering mechanism.

While the UK took the initiative for regulating production water and chemicals both nationally and under the Paris Convention, this has not been the case with the issue of drill cuttings. In fact, the UK made a reservation to the decision on muds and cuttings in 1988, because they were against the target of 100 g oil per 1 kg cuttings. Instead, they set a target of 150 g. The reasons behind this reservation were the lack of trust in the feasibility of a 100 g limit for the Southern North Sea and the lack of knowledge about the environmental effects of cleaning technology (United Kingdom, 1988). The UK lifted its reservation at the end of 1991 and has implemented the discharge limits for oily cuttings since then.

Besides the reporting requirements under the Paris Convention, the UK has also employed monitoring as a way to assess the impact of the offshore industry on the marine environment and, when possible, the effectiveness of reduction measures. For example, companies were required to monitor the sea bed because of discharges of mud and cuttings (Department of Trade and Industry, 2006c).

In Norway, concern and regulations for accidents and oil spills already existed since the 1960s, although operational pollution was not a target (yet) (Krosby, 1976). This concern for the marine environment is related to two factors: the fishing industry (Interviews Norwegian Oil Industry Association, 2006a; Statoil, 2006) and the fanatical devotion of Norwegians to outdoor recreation (Krosby, 1976). The protection of marine life against oil has been one of the drivers of the development of steering mechanisms, for example to prohibit the discharge of oil based muds (Interview Nederlandse Aardolie Maatschappij, 2005).

In the actor dimension, the Ministry of Petroleum and Energy is responsible for the offshore industry. This Ministry has the Norwegian Petroleum Directorate (NPD) that embodies this responsibility. Contrary to the UK and the Netherlands, however, it is not this Ministry or the NPD that is responsible for the environmental impacts of the offshore industry. Instead, the Norwegian Pollution Control Authority (SFT), established in 1974, regulates the environmental pollution of offshore oil and gas production. The SFT falls under the Ministry of the Environment.

In the actor dimension, the Norwegian oil industry is represented by the Norwegian Oil Industry Association (OLF). Contrary to the UK, the OLF was, until the mid-1980s, 'a weak reflection of its powerful British counterpart' (Nelsen, 1991, p. 145). Initially, the industry was not used to being regulated on environmental issues and were very reluctant to accept that; for example, the 40 ppm target was considered to be outrageous (Interview Hydro, 2006). During the 1980s, the Norwegian companies (Statoil, Hydro and Saga) adopted a more environmentally friendly attitude. According to Statoil (Interview Statoil, 2006), they did not want to become part of OLF, because the foreign companies with a much more reluctant attitude were the main members of OLF. Consequently, contacts between the industry, especially the Norwegian companies (Statoil, Hydro and Saga), and SFT took place on an individual basis (Interview Norwegian Pollution Control Authority, 2006a).

The first Norwegian steering mechanism focusing on reducing pollution did not emerge till 1981 when onshore pollution control was extended to cover offshore activities as well (Barrett & Howells, 1990). The Pollution Control Act 1981 requires platforms to have a discharge permit, because their activities cause pollution. The Act was accompanied with Guidelines about the application of discharge permits for the offshore industry in 1986 (Environmental & Safety Consultancy, 1992). However, as Statoil explains (Interview Statoil, 2006), until knowledge about the environmental effects was gathered and specific regulations were developed (e.g. the 40 ppm target), SFT could not do anything else then approve every discharge permit applied for. In later years, the discharge permits were used to implement the requirements set by the Paris decisions and recommendations.

During these years, the SFT was a very ambitious authority (Interview Hydro, 2006). For example, they wanted to set requirements, which were, according to the industry, too expensive or not effective (Interview Statoil, 2006). It was therefore in the interest of the industry to show SFT how the offshore industry operates and that its requirements were unfeasible. Because of that, a working relationship emerged in which the industry 'educated' SFT and SFT developed regulations based on that. This for example happened with the oil based muds when Statoil started to experiment with water based muds and shared their knowledge with SFT (Interview Statoil, 2006). Based on that, SFT could successfully pursue a restriction in the use of and a ban on the discharge of oil based muds both in Norway and under the Paris Convention.

An aspect of the compliance dimension in Norway is the reporting requirements for the offshore industry. These reporting requirements are also related to the reporting requirements that the Paris Convention poses onto Norway. The offshore industry has to report discharges of oil based muds and production water in an annual report. They also have to report the use of chemicals.

However, the Norwegian contribution to the compliance dimension of the regional sphere of authority has been the heavy focus it has always put on monitoring. The Norwegian authorities already required companies to report and monitor environmental conditions and effects around their platforms since 1973 (Gray *et al.*, 1999; Norwegian Oil Industry Association, 2007). Since SFT was a small authority with only a few people working on the environmental impacts of the offshore industry, the companies were the ones who were obliged to set up a monitoring program around their platforms (Interview Norwegian Pollution Control Authority, 2006a). According to Gray *et.al.* (1999), the SFT initially did not review the reports it received. This changed when SFT set up a system of expert review for the annual reports in the mid-1980s. At the same time, the SFT initiated the development of guidelines on how monitoring should be done to improve the quality of the reports. These guidelines were also brought forward within the Paris Commission, who adopted them in 1988. As Gray *et.al.* (1999) further explain, in 1995, a system of regional monitoring was introduced instead of the monitoring by individual companies. This is arranged through the OLF, although the actual monitoring is done by a consulting company. The monitoring activities in Norway were further expanded in 1999, when water column monitoring program was set up (Interviews Norwegian Oil Industry Association, 2006b, 2007).

In the Netherlands, the main governmental actor is the Ministry of Economic Affairs which is responsible for the offshore industry, including the environmental side of offshore oil and gas production. The State Supervision of Mines, which is part of the Ministry of Economic Affairs, is responsible for ensuring compliance from the offshore industry but is also an advisory body for its Ministry. The offshore industry is represented by the Netherlands Oil and Gas Exploration and Production Association (NOGEPA), which was established for that purpose in 1974.

National steering mechanisms were developed to implement Paris decisions and recommendations. In 1983, a chapter on the prevention of pollution by the offshore industry was added to the Mining Regulations Continental Shelf (Bus, 1993). These regulations prohibited the discharge of oil, oily mixtures, sanitary wastes or garbage of platforms. For all these discharges, except oil, the Minister of Economic Affairs is allowed to make exemptions subject to conditions. This was done through the 1987 Regulation for Discharge of Hydrocarbon Contaminated Mixtures from Mining Installations, which allows an average oil content of 40 ppm in production water and the initial 100 g oil per 1 kg of cuttings. The Netherlands is further more the only one without a specifically designed scheme through which the notification of chemicals used was arranged (Environmental & Safety Consultancy, 1992).

Deliberation between the Dutch government and the industry about the development of national steering mechanisms and the implementation of the regional ones took place within working groups hosted by NOGEPA. Working groups on monitoring, production water and toxic substances already existed in the 1980s. One example of deliberation that took place within these working groups is that the Dutch delegation would consult the industry before going to a meeting of the Paris Commission (Interview State Supervision of Mines, 2006). Although, at that time it was common that the government would indicate what had to happen and the industry would scarcely have a response to that (Interviews Gaz de France, 2007; State Supervision of Mines, 2006). At the same time, it was hard to get a grip on the offshore industry, because there was little information about the discharges from the offshore industry and the industry was reluctant to generate this information (Interview Ministry of Housing, Spatial Planning and the Environment, 2006).

In the compliance dimension, reporting is a mechanism that is mainly used for the oil content of production water. Reporting seems to be a less relevant compliance mechanism in the Netherlands than in the UK and Norway, since reporting on among others chemicals is lacking. Besides reporting, inspections on platforms were done regularly during the 1980s. Although it has to be noted that inspecting the 40 ppm limit is not easy, since it requires taking monsters and the oil content varies as well. Finally, monitoring has been initiated in 1983 to evaluate the effects of discharges of muds and cuttings (State Supervision of Mines, 2006).

5.3 The authority of the state during the 1970s and 1980s

The previous section showed that during the 1970s and mid-1980s, the main steering mechanisms in the environmental governance of the offshore industry were developed in the regional sphere of authority. More specific, the environmental governance of this industry is mainly based on restrictions and limits agreed under the Paris Convention. This section will first review the authority of the state in developing these standards and limits under the Paris Convention. To implement the Paris standards, national frameworks were developed within the UK, Norway and the Netherlands separately. Section 5.3.2 will therefore discuss the authority of the state in generating compliance within the different national contexts.

5.3.1 Authority in developing steering mechanisms

The development of steering mechanisms that deal with pollution from the offshore industry has known a relatively slow start. Where oil pollution from shipping was already an issue in the 1950s and 1960s and environmental issues hit the international agenda in the early 1970s, oil pollution from the offshore industry did not receive attention till the late 1970s. According to Rowan-Robinson (2000), this may be due to the belief that the offshore industry was not a significant polluter and that the flushing capacity of the North Sea would level off pollution. This belief was also not threatened by any environmental accidents or incidents. There was an incident on the Norwegian platform Ekofisk Bravo in 1977, when a well was not closed properly and blew out. This resulted in a major oil spill in the North Sea. However, 30-40% of the oil evaporated and, according to the Norwegian State Pollution Control Board, no major ecological damage was done (IncidentNews, 1977).

The development of steering mechanisms has been taking place at the regional level, i.e. under the Paris Convention, from the beginning. However, it is the national governmental actors who have pursued the development of regional steering mechanisms. For example, we have seen that both the UK and Norway have initiated decisions or recommendations on production water and oil based muds respectively. Besides the UK and Norway, the Netherlands is active in the development of regional steering mechanisms as well. For example, these three states together submit the majority of the documents discussed within the GOP meetings (see GOP summary records 1984 - 1994).

It is no coincidence that these three states are most active and have most authority over the steering mechanisms adopted, because they together account for the vast majority of the platforms in the North-east Atlantic. These three states are thus the most powerful ones within the Paris Convention when it comes to developing steering mechanisms for the offshore industry.

The North Sea Ministerial Conferences have contributed to the authority of the UK, Norway and the Netherlands over the development of steering mechanisms. The NSMCs were a forum with a higher political profile and with fewer members. For the UK, Norway and the Netherlands this has meant an extra opportunity to come to consensus and to give political impetus to the decision-making processes under the Paris Convention.

But the industry, through their national and international industry associations, has been aiming to gain authority the development of steering mechanisms as well. Although the rules of the game of the regional sphere of authority did not allow direct participation in the main forums of decision making (i.e. the GOP and Paris Commission) there were other ways through which they could influence what was negotiated in these forums. First of all by lobbying the national delegations. In for example the UK, a fairly close relationship between the government and industry existed (Interview Department of Trade and Industry, 2006c). In Norway, an important part of the industry was a state-owned company, i.e. Statoil. Second, during the 1980s, the E & P Forum and the GOP discussed specific issues during specially organized meetings. This was predominantly done to discuss the issue of oil based muds and cuttings. Finally, the international industry associations were allowed to give presentations at the beginning of GOP meetings, an opportunity that the E & P Forum, and later also EOSCA, have regularly used.

The development of steering mechanisms was thus something between the governmental and industrial actors, because other societal actors, i.e. environmental NGOs, were not part of this sphere of authority. It is likely, that the authority of the industry, sometimes via the governments, has been considerable, because there was no competition between the industry and environmental NGOs and because the relations between governmental actors and the industry were close. The latter, also makes it difficult to make more specific observations about the exact authority and influence of the industry and governmental actors, especially because it would require a more in-depth analysis of the environmental governance of offshore oil and gas production than is done in this thesis.

However, it should also be noted that before the industrial actors were ready to accept that discharges of their exploration and production activities could harm the environment, they had to overcome their reluctance. Initially, the industry was reluctant to deal with environmental issues, because until then, they had never been confronted with these issues (Interview Hydro, 2006). According to Gray *et.al.* (1999), the industry even had a hostile reception of the suggestion that ecological effects extended over a larger area than originally assumed.

Remarkably, the claims about such biological effects were based on monitoring reports of the industry itself (Gray *et al.*, 1999). Both the UK and Norway required the offshore industry to monitor the environmental conditions around their platforms (Department of Trade and Industry, 2006c; Gray *et al.*, 1999). Besides monitoring studies, both the industry and the government initiated studies to study environmental effects from discharges of oil based muds, cuttings and chemicals (Secretariat of the Paris Commission, 1986; Taylor, 1991). Sharing of the results of these studies was among others done through the meetings between the GOP and the E & P Forum.

All in all, the main authority of the industry lied in generating information about their industry activities, discharges and environmental effects and to lobby to get their views integrated into the steering mechanisms. The governmental actors displayed authority by negotiating steering mechanisms among others based on the information received by the industry. In addition, the governmental actors were had a much more active role in the development of steering mechanisms, while the industry was more passive (State Supervision of Mines, 2006).

5.3.2 Authority in generating compliance

During the early 1980s, compliance mechanisms were not well-developed yet. Initially, there was also no need for them, because the steering mechanisms did not have any mandatory reduction features, i.e. the 40 ppm for produced water was a provisional target and the chemical notification scheme was about gathering information about the use of chemicals offshore.

This changed when the 40 ppm became a real target and when the discharge of oil based muds and cuttings was banned or restricted. As explained above, compliance with the 40 ppm target was based on sampling and reporting to the governmental authorities. These governmental authorities in turn had to report to the GOP and Paris Commission. While non-compliance of installations was discussed during the GOP meetings, sanctions were never imposed (see GOP summary records 1984-1994).

Ensuring compliance with the discharge limits of oil based muds and cuttings was also based on reporting requirements. The annual report on the discharges from offshore installations compiled the information on oil discharges through produced water, oil based muds and oil cuttings. Compliance and non-compliance was therefore very transparent for all members of the GOP. Non-compliance with the ban on oil based muds and oily cuttings has, as far as the summary records and annual reports show, not been an issue.

A compliance mechanism that is related to the permits and licenses, that the offshore industry need to have to be able to explore and produce oil, is inspections. During the 1980s, the UK, Norway and the Netherlands employed a few inspectors, who inspected and audited platforms. However, the question is whether these inspections were focused on the environmental requirements of the Paris steering mechanisms or more onto other areas such as safety or working conditions. Moreover, it has to be kept in mind that enforcing standards such as the 40 ppm target through inspections is difficult to maintain, because it is an average and not a maximum (Environmental & Safety Consultancy, 1992). A practical issue that exists with regard to these inspections is that inspections cannot be done unannounced (Barrett & Howells, 1990; Environmental & Safety Consultancy, 1992; Rowan-Robinson, 2000). First, because inspectors rely on transportation provided by the operator of the platform. Second, because at least the UK and Norway have prohibited unannounced landings on installations (Barrett & Howells, 1990).

A second activity that relates to the compliance dimension is monitoring. Although monitoring is not directly focused on ensuring compliance it does help to: (i) assess whether measures are effective and (ii) to assess the impact of activities on the marine environment. The first helps in evaluating the existing measures, while the second can help to detect issues that need to be addressed. This difference also comes back in the monitoring activities by the offshore industry. For example, sea-bed monitoring was used to generate insight in the effects of muds and cuttings and the reduction in impacts of these discharges after the ban in discharges was in place.

It can be concluded that in the regional environmental governance of the offshore industry in the 1980s and early 1990s, ensuring compliance was mainly based on reporting. The governmental actors were responsible for ensuring compliance and for providing the Paris Secretariat with the required monitoring information. However, in doing so the governmental actors rely quite heavily on the industry to monitor and report their discharges (Rowan-Robinson, 2000). Thus again, the role of the industry was to provide information and knowledge, while the governmental actors use this knowledge to judge the level of compliance and intervene accordingly.

5.4 The emerging national spheres of authority during the 1990s and 2000s

As discussed above, until the mid-1990s, national actors participated in the regional sphere of authority and the regional steering mechanisms were implemented in the UK, Norway and the Netherlands. However, since the mid-1990s, national spheres of authority developed within the UK, Norway and the Netherlands besides the already existing regional sphere of authority. Especially Norway and the Netherlands have developed their own approach towards preventing environmental pollution from the offshore industry, going beyond a mere implementation of Paris/OSPAR decisions and recommendations. On top of that, in all three states, the relationship between the offshore industry and the national governmental actors developed into a more cooperative relationship as a result of joint initiatives. The existence of and the interaction between the national and regional spheres of authority in the environmental governance of oil and gas production is visualized in Figure 5.

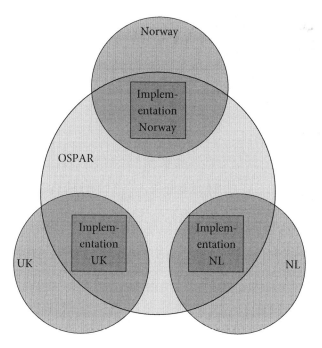

Figure 5. The regional and national spheres of authority.

5.4.1 United Kingdom

During the discussion of the implementation and compliance of regional steering mechanisms by the UK in the previous section, it became clear that the main national actors are the Department of Trade and Industry as a governmental actor and UKOOA as in industry actor. The national steering mechanisms developed until the mid 1990s are the Prevention of Oil Pollution Act of 1971 and the voluntary Offshore Chemical Notification Scheme. Moreover, compliance is ensured through monitoring and reporting.

The emergence of a national sphere of authority in the UK is, first of all, related to the establishment of the Oil and Gas Industry Task Force (OGITF) by the Secretary of State for Trade and Industry at the end of 1998. The task force aimed to recommend ways in which the UK oil and gas industry could maintain its competitiveness in the face of increasing maturity of the oil and gas fields and low oil prices (Oil & Gas Industry Task Force, 1999). One of the working groups specifically focused on the environmental and sustainability trends facing the industry in the future. Actors involved in this taskforce were DTI, Department of Environment, Transport and the Regions, UKOOA and individual oil companies.

The first outcome of this OGITF was a set of environmental principles that should guide the development of environmental regulation in the future (Oil & Gas Industry Task Force, 1999). These principles were continual improvement through long term goal setting, flexible methods which promote innovation, sound scientific analysis and a precautionary approach, balance between cost and benefit and tradeoffs between emission streams, a transparent and participatory decision making process and the application of penalties. In a way, these principles show an agreed discourse about which criteria future steering mechanisms should meet.

The second outcome was the recognition that a more effective dialogue between the government, the industry and the environmental NGOs was needed (Oil & Gas Industry Task Force, 1999). One of the lessons of Brent Spar therefore gained momentum within the OGITF. In 1995, the offshore industry and the UK government were astonished by the international public outcry against the sinking of the old storage buoy Brent Spar. It was the first time that environmental NGOs and the public were making demands from the offshore industry. Even though the Brent Spar is related to decommissioning issues, it showed to both the industry and the governmental authorities that there was a need for a clear environmental regime, which should also be communicated to the public (Interviews Consultant offshore industry, 2006; Department of Trade and Industry, 2006a; Vanner, 2006). Or as UKOOA (2001, p. 38) puts it: 'compliance alone was insufficient; stakeholder engagement was recognized as an imperative in major environmental issues. Since then, UKOOA and its member companies have worked with NGOs to build communications links and facilitate dialogue'.

This recognition shows a new discourse that emphasizes the need of stakeholder engagement and dialogue. But the outcome was not only a new discourse but also two stakeholder forums: an industry-government and a NGO-industry-government forum. Both forums ensure that interactions between the government, industry and NGOs take place on a regular basis. The meeting with NGOs (the so-called UK Offshore Forum), give the government and industry the opportunity to inform the NGOs about what is going on, while it presents NGOs with the opportunity to ask questions and to express their concerns (Interview British Petroleum, 2006).

The government-industry forum creates mutual understanding between the government and the industry. It is also one of the places where OSPAR developments are discussed (Interview British Petroleum, 2006).

The new discourse and UK Offshore Forum also mean that environmental NGOs became part of the actor dimension of the national sphere of authority. So far, environmental NGOs had not been in the picture of both DTI and the offshore industry (Interview Department of Trade and Industry, 2006c). They were for example not involved in the task force itself. From 1999 onwards, the government, industry and NGOs met regularly through the UK Offshore Forum. Environmental NGOs that participated in this forum have among others been: World Wildlife Fund, Royal Society for the Protection of Birds, Whale and Dolphin Conservation Society and the Marine Conservation Society.

The third outcome of OGITF was the plan to develop a new steering mechanism with regard to gas flaring (Oil & Gas Industry Task Force, 1999). Gas flaring is a process in which gas is burned in cases of unplanned over-pressure. Since 1998, the Petroleum Act requires each platform to have a permit for flaring gas. Between 1990 and 2001, a reduction of 30% in flaring practices was already achieved. However, further measures were deemed necessary to reduce the amount of gas flared even further (Flare Transfer Pilot Trading Scheme Steering Committee, 2002). It was decided to set up a Flare Transfer Pilot Trading Scheme. This trading scheme would on the one hand serve as a pilot for the concept of emissions trading, but would at the same time aim to reduce emissions from gas flaring. The Trading Scheme is a joint government-industry initiative and more reduction has been achieved than projected (Flare Transfer Pilot Trading Scheme Steering Committee, 2006). In 2005, the decision was taken to integrate flare trading within the European Union Emissions Trading Scheme as of 2008 (Flare Transfer Pilot Trading Scheme Steering Committee, 2006).

The fourth outcome of OGITF was the UKOOA sustainability strategy (Oil & Gas Industry Task Force, 1999). The development of sustainability strategies was initiated by a cross-sectoral government initiative to develop such strategies. UKOOA developed the one for the offshore industry in 2001.

The Sustainability Strategy of the offshore industry is an industry steering mechanism and contains a Sustainable Development Action Plan with regard to environmental management (United Kingdom Offshore Operators Association, 2001). In this action plan, the industry commits itself to continual improvement in the industry's environmental performance, research commitments with regard to environmentally sensitive areas, the sharing of best practices, education of staff, etc. One element that came back in this strategy was the development of environmental management systems, i.e. the aim was to have 90% of the UK oil and gas production covered by an independently verified environmental management system by the end of 2001 and all oil and gas production by the end of 2002. Other aims include 30 ppm of oil in produced water and 30% less oil spilled by the end of 2003.

Compliance with this strategy is assessed every year in an annual report. Moreover, in 2003, 2005 and 2007, the strategy was reviewed. Each of these reviews was preceded by stakeholder events to discuss how the strategy and indicators used can be improved. Here the discourse of stakeholder dialogue comes back, because UKOOA sees these stakeholder events as to 'complement discussions with NGOs at the UK Offshore Forum, as well as other consultations on specific issues when they arise' (Oil & Gas UK, 2007, p. 3). With regard to goals set, it can be noted that the strategy in 2001 initiated several goals that went beyond existing regional and national steering mechanisms. In contrast, in 2007, the goals set are either in line with regional and national steering mechanisms or focus on exploring new issues such as noise and carbon storage (Oil & Gas UK, 2007).

Besides the outcomes that OGITF produced in terms of new steering mechanisms and new structures for deliberation between the government and the industry, an important change in the rules of the game contributed to the development of a national sphere of authority as well. In the late 1990s, DTI was granted with powers to develop secondary legislation (Interviews British Petroleum, 2006; Department of Trade and Industry, 2006a). The UK has a legislative system in which primary legislation refers to main laws that are passed by the legislative bodies of the UK. Secondary legislation is delegated legislation that does not have to pass the legislative bodies. As DTI explains (Interview Department of Trade and Industry, 2006a), this delegated legislative power means that DTI is able to develop standards and regulations without having to go through the time-consuming process of passing legislation through Parliament.

The common practice to implement OSPAR regulations through voluntary agreements with the industry (Interview Department of Trade and Industry, 2006b) could therefore be replaced by formal regulations. For example, the voluntary Offshore Chemical Notification Scheme was replaced by the Offshore Chemicals Regulations in 2002. These regulations implemented the OSPAR chemical scheme that had been evolving since 1996. But regulations to regulate oil pollution from the offshore industry have been renewed as well. The 1971 Prevention of Oil Pollution Act is still in force, but DTI adopted a separate regulation that is more suitable for the offshore industry in 2005, i.e. the Offshore Petroleum Activities (Oil Pollution Prevention and Control) Regulations.

The joint initiative of OGITF, the new discourse on stakeholder dialogue and the joint initiatives in terms of steering mechanisms show that the power dimension is not necessarily shaped by a regulator-complier relationship. On the one hand, DTI clearly regulates the industry and has the power to develop formal regulations while the industry is subject to these formal regulations and has to comply. What is more, the new formal regulations require the offshore industry to have permits for oil discharges, chemicals and flaring. The regulator-complier relationship is therefore shaped by the application of permits, operating under permits and inspections to check compliance.

On the other hand, specific initiatives have shifted some responsibility and power to influence steering mechanisms to the industry. For example, the principles for future environmental regulations have enabled the industry indirectly to shape the characteristics of future regulation. The forums in which dialogue takes place between the government and the industry also ensure input from the industry in the development of steering mechanisms. Finally, the industry has taken matters more in their own hands through the development of an industry sustainability strategy.

5.4.2 Norway

When discussing the Norwegian implementation of the regional steering mechanisms above, some elements of the Norwegian national sphere of authority were already revealed. For example, the main governmental actors in Norway are the SFT and NPD. In environmental matters, SFT regulates the discharges to water, while NPD is responsible for air emissions. Moreover, as will become clear below, by the mid-1990s, OLF had become a much stronger representative for the offshore industry in Norway than it was during the 1980s.

As also discussed, the main steering mechanism in Norway was the requirement of a discharge permit through the 1981 Pollution Control Act. Reduction measures, such as the 40 ppm target, are part of this permit. In the 1980s and early 1990s, this steering mechanism was used to implement and ensure compliance with standards and regulations coming from the Paris Convention. This permit system continues to be the backbone of the Norwegian environmental governance of the offshore industry (Interviews Hydro, 2006; Norwegian Pollution Control Authority, 2006b).

The first step that marks the development of a national sphere of authority which does more than only implementing regional regulations is the adoption of a CO_2-tax by the NPD in December 1990. The tax applies to petroleum burned and CO_2 emissions emitted offshore. With the CO_2 tax, the NPD aims to have an economic incentive to stimulate operators to burn less petroleum and to reduce CO_2 emissions. Initially, the offshore industry indeed made investments that led to less CO_2 emissions. This, however, only works as long as the CO_2-tax is higher then the investments needed to reduce CO_2, which is no longer the case (Interview Norwegian Oil Industry Association, 2006a).

The second step in the development of the Norwegian national sphere of authority is the initiative of the Norwegian government to develop a consultative forum, called Miljøsok, in 1995. In Miljøsok, the government and, among others, OLF debated how to ensure that the offshore industry remained in the lead of environmental-friendly oil and gas production (Norwegian Petroleum Directorate, 2000). Reduction of CO_2 emissions was one of the main issues in this regard. The outcome of the first phase (in December 1996) was an overview of the environmental issues and a set of objectives, targets and recommendations for the offshore industry and the government.

The second phase, which ran between 1997 and 2000, focused on the implementation of these recommendations. The outcome of this phase is more ambiguous, since by that time, realization grew that especially the objective of stabilizing CO_2 emissions was unrealistic (Norwegian Petroleum Directorate, 2000). In 2001, the Miljøsok was followed-up by the Miljøforum. Miljøforum was developed to have a permanent forum to preserve the dialogue and mutual understanding between the government, the offshore and fishing industry and the environmental NGOs that had emerged through Miljøsok (Interview Norwegian Oil Industry Association, 2006b).

Parallel to Miljøsok and -forum, a second steering mechanism, the zero harmful discharge approach, was developed. In 1996, the Ministry of Environment initiated a zero harmful discharge philosophy for the offshore industry. The idea of the zero harmful discharge philosophy was launched in the White Paper 'Environmental Policy for a Sustainable Development' of the Ministry of Environment. This White Paper established two principles (1) to not exceed the levels of critical loads on ecosystems (i.e. nature's carrying capacity) and (2) to act in a precautionary way (Ministry of Environment (Norway), 1996-97).

While the Norwegian government initiated this new approach, the responsibility for having zero harmful discharges lies with the operator, who has to identify and evaluate measures for its installations. The development of this policy is characterized by periods in which SFT takes the initiative, which are subsequently alternated with periods in which the policy is elaborated in government-industry working groups.

For example, after the White paper, a Zero Discharge Group – in which the Norwegian authorities, oil companies and supply industry participated – started late 1997 (Marthinsen & Sørgård, 2002). The report of the Zero Discharge Group discussed the use of concepts and terms and provided an overview and recommendations for further work (Zero Discharge Group, 2003).

After that, the SFT took the initiative by ordering the operators to develop field specific plans to achieve the zero harmful discharge target by 2005. By 2000, the operators had developed these plans and progress is reported to SFT annually (Nilssen & Øren, 2003). Most of the work focuses on reducing oil discharges through produced water, discharges of hazardous chemicals and discharges of cuttings. In 2002 and 2003, the Zero Discharge Group resumed its work and was asked to review the targets and definitions, to develop a format for reporting and to review the development of technology (Zero Discharge Group, 2003). After that, however, progress has been limited. The objective was to have zero harmful discharges by 2005. This target was, however, not reached due to delays in the implementation of measures (Interviews Norwegian Oil Industry Association, 2006b; Norwegian Pollution Control Authority, 2006b). A follow-up of the zero harmful discharge work has also not been made public by the SFT yet.

As many of the interviewees noted, it is important to understand that the objective of zero discharges is not about bringing the actual discharges to zero, but about making sure that the discharges have zero harmful impact on the environment (Interviews Norwegian Oil Industry Association, 2006b; Norwegian Pollution Control Authority, 2006b). That is why it is in fact a zero *harmful* discharge approach. This for the offshore industry important nuance was also the outcome of the Zero Discharge Group in 1998 (Marthinsen & Sørgård, 2002). Adding the term harmful to the zero discharge approach is thus an important discursive shift. First, because adding the notion harmful has made it an approach based on risk rather than the precautionary approach of zero discharges, i.e. a ban on all discharges (Interviews Hydro, 2006; Norwegian Oil Industry Association, 2006b; Norwegian Pollution Control Authority, 2006b). What matters is not the discharge itself, but the risk that the discharge presents harm to marine life and ecosystems. It is this risk that needs to be brought to zero.

Second, this risk-based approach is associated with so-called functional requirements (Interview Norwegian Pollution Control Authority, 2006b). Functional requirements are focused on the function of a system or operations and do not necessarily state how this function needs to be achieved. This provides the offshore industry with flexibility in how to design their systems and operations.

The discourse of risk-based approaches and functional requirements also comes back in another steering mechanism which was adopted in 2001: the Health, Safety and Environment in the Petroleum Activities Regulations (HSE Regulations). These regulations set down functional requirements with regard to minimizing discharges and emissions to the external environment (Petroleum Safety Authority Norway, 2008). The HSE Regulations integrated the previous separate regimes around safety, health and environment (Interview Norwegian Oil Industry Association, 2006b).

However, the HSE Regulations also contain prescriptive requirements, which do not present the offshore industry with new standards and requirements. The main prescriptive environmental requirements in the HSE Regulations are those coming from the Paris/OSPAR Commission, i.e. the 40 ppm target and the harmonized mandatory control system for chemicals. The difference is that it is no longer necessary to put these requirements in all individual discharge permits, because through the HSE regulations they apply to the whole offshore industry (Interview Norwegian Pollution Control Authority, 2006a).

Similar to the UK sphere of authority, there were several initiatives (zero discharge working group, Miljøsok and -forum) that resulted in a discourse around stakeholder dialogue. In Norway, however, the emphasis lies on mutual understanding between the government and the offshore industry, but also between the offshore industry and the fishing industry. The concern with fishing is not new, but regular dialogue between the offshore industry and the fishing industry was enhanced through Miljøsok.

In addition, similar to what happened in the UK, environmental NGOs became part of the actor dimension of the sphere of authority through Miljøsok and -forum. However, according to the industry association (Interview Norwegian Oil Industry Association, 2006b), NGO participation in Miljøforum has gone up and down. One of the main environmental NGOs that focuses on the offshore industry, Bellona, is especially active during licensing rounds and on the Barents Sea (Interviews Bellona, 2006; Norwegian Oil Industry Association, 2006b). This however, does not fit well with the purposes of the Miljøforum, which focuses much more on operational pollution.

The permit system, HSE Regulations, Miljøsok and -forum, and the zero harmful discharge approach have resulted in a power dimension that displays two kinds of relationships between the government and the industry. The discharge permit system and the HSE Regulations through which the Norwegian government regulates environmental pollution from the offshore industry puts the government and industry in a regulator-complier relationship where the power to regulate rests with the government.

However, the already cooperative relationship that existed with regard to monitoring practices has now been extended further. First through the initiative of Miljøsok, a shared understanding about environmental issues, feasible targets and possible measures was developed. Under the zero harmful discharge approach, the government and industry have worked together in developing the approach and monitoring progress. This shows that it is no longer the government alone who sets standards, but that the industry is influencing the development of steering mechanisms as well. What is more, because of the turn to functional requirements, the industry has a lot of decisive power in how certain targets will be reached.

It has to be noted however, that the influence of the industry is subject to checks-and-balances. For example, the individual zero harmful discharge plans had to be approved by the SFT. In addition, SFT will take further action, when the industry is not able to achieve the zero harmful discharge objective under the current scheme (Nilssen & Øren, 2003; Interviews Norwegian Oil Industry Association, 2006b; Norwegian Pollution Control Authority, 2006b).

In the compliance dimension, reporting continues to be an important tool for evaluating operators' work in reducing the environmental impact from the operations (Nilssen & Øren, 2003). Norwegian operators are required to submit a field-specific annual report about among others their zero-discharge work and the actual discharges and emissions of that year (Nilssen & Øren, 2003).

Environmental monitoring also remains an important compliance mechanism. The monitoring efforts have even been extended in recent years. While monitoring the seabed has been common practice since 1973, water column monitoring was introduced in 1999 (Interview Norwegian Oil Industry Association, 2006b; Norwegian Oil Industry Association, 2007). The heavy emphasis that lies on monitoring and the shift to water column monitoring within the Norwegian sphere of authority is consistent with the risk-based approach that has become important since the mid-1990s.

The last compliance mechanism that the Norwegian government uses is inspecting and auditing platforms. These audits have become more comprehensive now that a whole range of steering mechanisms exist. So when the SFT audits an oil company, it does that according to the discharge permit, the HSE regulations, but also the internal documents within the industry (Interview Norwegian Pollution Control Authority, 2006a).

5.4.3 The Netherlands

The section on the implementation and compliance of regional steering mechanisms in the Netherlands above shows that the Ministry of Economic Affairs together with its State Supervision of Mines were the main governmental actors active in the production of oil and gas. The Dutch industry association is NOGEPA. Deliberation between these actors takes place in working groups. The main national steering mechanism is the set of Mining Regulations. Finally, reporting is the main mechanism through which compliance is ensured.

Instrumental in the development of the Dutch national sphere of authority is the 'Declaration of intent for the implementation of environmental policy for the oil and gas industry' (hereafter referred to as the Environmental Covenant) signed by the Dutch government and the offshore industry in 1995. The Dutch government was represented by the Minister of Economic Affairs, the Minister of Housing, Physical Planning and the Environment, and the Minister of Transport and Public Works. The industry was represented by the NOGEPA and the individual companies that are active in oil and gas exploration and production in the Netherlands, i.e. both onshore and on the Dutch Continental Shelf. In 2001, 'Facilitair Organisatie Industrie' (FO-Industry) was asked to participate in the covenant to fulfil many of the administrative tasks, such as writing the guidelines and annual reports. Since FO-Industry does this for many Environmental Covenants, they facilitate the Environmental Covenant for the offshore industry with their experiences from other covenants (Interview FO-Industry, 2006). These actors together shape the actor dimension of this national sphere of authority.

The Environmental Covenant finds its origin in the 'target group policy' initiated in the first National Environmental Policy Plan of the Dutch government, published in 1989 (Ministerie van Volkshuisvesting; Ruimtelijke Ordening en Milieubeheer, Ministerie van Economische Zaken, Ministerie van Landbouw en Visserij & Ministerie van Verkeer en Waterstaat, 1989). The plan focused on the responsibility of companies and industries in reaching the environmental targets set. That is why the government initiated a target group policy in which consultation and deliberation structures with the target groups of the policies would be created.

Consequently, along with the target group policy a new discourse on industry responsibility and dialogue was expressed. The idea behind this discourse is that the government is not able to develop and implement effective environmental policies on its own. Large investments and efforts from private parties would be needed to intensify environmental policy and to achieve sustainable development (Ministerie van Volkshuisvesting; Ruimtelijke Ordening en Milieubeheer *et al.*, 1989).

There were several reasons why the government was interested in adopting an Environmental Covenant for oil and gas production. First, because the offshore industry operates outside territorial waters, the oil and gas industry was not covered by a set of environmental standards like other industries were; the Environmental Covenant could provide such a set of environmental standards

(Interview Wintershall Noordzee B.V., 2006). A second reason was to improve the grip or control of the government over the offshore industry. One of the governmental actors explains that controlling this industry was difficult because everything happens in a big pool of water, but also because the companies were reluctant in making an effort to study their own discharges and share the results with the government (Interview Ministry of Housing, Spatial Planning and the Environment, 2006).

For the oil and gas companies, an important reason to participate in the covenant was to avoid upcoming environmental legislation. The Environmental Covenant provided for a trade off in which the industry would commit itself to reduction targets and the government would not impose environmental legislation on the industry (Interview Wintershall Noordzee B.V., 2006). Another reason why the Environmental Covenant was attractive is the clear policy it provides for the long term (Interview FO-Industry, 2006). This would facilitate the decision making around long term investments and transformations in the production process. In addition, the Environmental Covenant allowed more flexibility in the time-path for the achievement of the targets and the way in which these targets would be achieved (Interview FO-Industry, 2006).

The Covenant describes the Integrated Environmental Target Plan for the oil and gas industry onshore and offshore, which should be achieved in 2000 and 2010. These targets focus on climate change, acidification, heavy metals and benzene, oil, added chemicals, waste and environmental care. Even though these targets are based on government policies, during the development of the Environmental Covenant negotiations between the government and industry have among others focused on the extent to which these goals were feasible for the offshore industry (Interviews FO-Industry, 2006; Institute for Marine Resources and Ecosystem Studies, 2006). However, according to the State Supervision of Mines (Interview State Supervision of Mines, 2006), there was little room for negotiation, because the Environmental Covenants with other industries meant that a rough blueprint for the approach towards Environmental Covenants was already laid down. Nevertheless, through this Environmental Covenant the industry has actively committed itself to these goals.

De Jong and Henriquez (2000) explain the procedures in the rules of the game dimension that help to implement the Environmental Covenant. To achieve the targets, each company has to develop a 4-year Company Environmental Plan. These plans are discussed in the working groups under the Environmental Covenant and have to be approved by the Ministry of Economic Affairs. An aggregated Industry Environmental Plan is developed by NOGEPA based on the individual Company Plans and has to be approved by the Ministry as well. Progress about the implementation of the Environmental plans is submitted and discussed annually. It should be noted that as a result of these rules of the game, the reporting requirements to ensure compliance have expanded as well, i.e. reporting is not only a compliance mechanism for OSPAR decisions and recommendations, but also for the Environmental Covenant.

While the development and implementation of the Environmental Covenant is mainly a government-industry matter, there were some occasions where Dutch environmental NGOs have voiced their opinion about the Declaration of Intent itself or about the Environmental Plans. For example, they expressed their concerns about the future effectiveness of the Environmental Covenant or they used consultation rounds to comment the Industry or Company Environmental Plans (based on documents of the archive of the North Sea Foundation). However, there is no structural contact with environmental NGOs (Interview Wintershall Noordzee B.V., 2006). Environmental NGOs are therefore not a structural part of the actor dimension, e.g. they are not participating in the Environmental Covenant, but try to influence the sphere of authority on an ad hoc basis.

The Environmental Covenant is also important for the structured deliberation it provides and the mutual understanding it creates between the government and the industry (Interviews FO-Industry, 2006; State Supervision of Mines, 2006). This structured deliberation is not only used for implementing the targets of the Environmental Covenant, but also for the development of other environmental policies that might influence the target plan of the covenant (Ministerie van Economische Zaken & Ministerie van Volkshuisvesting; Ruimtelijke Ordening en Milieubeheer, 1996).

For example, the Environmental Covenant serves as a forum to discuss the development within OSPAR and the implementation of OSPAR decisions and recommendations (Interviews FO-Industry, 2006; Ministry of Transport, Public Works and Water Management, 2006; State Supervision of Mines, 2006). The close link between the Environmental Covenant and OSPAR also come back in a change in working groups under the Environmental Covenant. In 2002, the working groups 'OSPAR implementation', 'drill cuttings' and 'chemicals' (to implement the OSPAR Harmonized Mandatory Control Scheme) were set up (Netherlands Oil and Gas Exploration and Production Association, 2002). Moreover, in the evaluation of the Covenant in 2001, the industry indicated that they attach great importance to discuss new issues, such as emissions trading, benchmarking, etc., within the deliberation structure of the Covenant (FO-Industrie & PricewaterhouseCoopers N.V., 2001). In this way, an integral approach for all environmental issues can be ensured.

Another development in the steering mechanism dimension is the adoption of a new Mining Act, Mining Regulations and Mining Decree in 2003. The new Mining laws were developed because the old mining laws were no longer appropriate to the situation at the end of the 20th Century. For example, the new mining laws have simplified and integrated all separate mining laws, regulations and ordinances that existed before.

The main new element of this Mining Act is the requirement of an Environmental Mining permit for those installations that do not require an environmental permit under the Environmental Protection Act (Ministerie van Economische Zaken, 2003). Since the Environmental Protection Act only applies to Dutch territory, thus till 12 nm out of the coast, all installations in the Dutch Economic Exclusive Zone are required to get this Environmental Mining permit. The Mining Decree repeats the prohibition of the discharge of oil, mixtures containing oil, sanitary garbage and garbage from installations and the exemptions to this rule. The Mining Regulations are used to implement the specific requirements that come from OSPAR decisions or recommendations, e.g. the 40 ppm oil in produced water norm, the ban on the discharge of oil based muds, but also the whole chemical notification scheme.

Similar to the UK and Norway, the power dimension of the Dutch sphere of authority is still characterized by the conventional regulator-complier relationship between the government and the industry. Initially this relationship was not characterized by the requirements of permits, but by standards and requirements set through the Mining Regulations. Since the 2003 Mining Law, all platforms need to have an environmental permit. Permit application and inspections have therefore become characteristic in this regulator-complier relationship.

On the other hand, a cooperative relationship with shared responsibilities between the government and industry exists through the Environmental Covenant. Since this covenant is the main source of environmental standards, this relationship puts a bigger mark on the power dimension of the Dutch sphere of authority than the Mining Laws. However, it should be noted that similar to the Norwegian power dimension, there are some checks and balances within the Environmental Covenant. At these moments, it is the government that takes the final decision, for example in approving the industry environmental plans.

5.5 The emergence of industrial steering mechanisms during the 1990s and 2000s

The period between the mid-1990s and 2000s, was not only marked by the emergences of national spheres of authority, but also by the development of environmental steering and compliance mechanisms within companies and the industry itself. This is expressed through the development of company environmental policies, codes of practices, environmental management systems and industry standards. While it goes too far to analyse environmental policies and strategies of individual companies (see Van de Wateringen, 2005 for that), the emergence of industry steering and compliance mechanism is a trend that has taken place across the whole industry.

Yet, it goes too far to denote the emergence of industry steering mechanisms as taking place in a new sphere of authority. A sphere of authority refers to an arrangement of actors that interact with each other in developing and implementing steering mechanisms within a certain institutional context. I argue that this institutional context is still lacking or at best too fragmented to regard it as a single sphere of authority. For example, the International Organization for Standardization (ISO) and its ISO 14001 standards can be considered a sphere of authority, but while some individual oil companies are part of this ISO sphere of authority others are not (because they implement a different kind of environmental management system). The lack of a coherent institutional framework and steering mechanisms provided for by the industrial associations on the national and international level (i.e. OLF, NOGEPA, UKOOA and E & P Forum) makes the development of industry 'self-governance' too fragmented.

5.5.1 Company environmental policies and environmental management systems

The attitude of the offshore industry towards environmental issues and environmental regulations improved in the beginning of the 1990s. The environmental consciousness of the companies and the people within the companies started to change, because they recognized that environmental regulation was both necessary and feasible. According to Van de Wateringen (2005), during this period the impact of environmental issues on the oil industry became apparent through the oil spill by the Exxon Valdez in 1989, the Brent Spar incident en the execution of Ken Saro Wiwa in 1995. For example, until then, the offshore industry had never fully accepted the 40 ppm limit for oil in produced water (Interview State Supervision of Mines, 2006). In addition, companies realized that better environmental performance and showing that they performed better was beneficial to them (Interview Norwegian Oil Industry Association, 2006b). For Shell in particular, the Brent Spar incident was the trigger to develop an external focus instead of continuing their internal focus (Van de Wateringen, 2005). Furthermore, Shell started to improve the integration of environmental issues in their management system after Brent Spar.

Subsequently, companies started to develop environmental codes of practice, environmental statements and policies, and environmental management systems. As Van de Wateringen (2005) observes, a substantial part of the petroleum companies had already issued their first environmental report before 1996 and many followed between 1996 and 1999. More importantly, almost all environmental reports – both the first and later ones – mention the existence of a corporate environmental policy.

The use of management systems as corporate steering mechanisms is not new. It became a common phenomenon in the offshore industry at the end of the 1980s. But, these management systems focused on safety only. In the course of the 1990s, however, environmental management systems were developed as well. Some companies already had an environmental management system before 1996, while others followed in the second half of the 1990s (Van de Wateringen, 2005).

This, of course, does not mean that the industry was not active at al in environmental issues before they started to develop and report their environmental management strategies. Statoil, for example, already experimented with the less polluting water based muds instead of oil based muds in the 1980s (Interview Statoil, 2006). It also tried to find information about the hazardousness of certain chemicals that it used. BP decided to inject drilling muds and produced water on some platforms in the very early 1990s, which was well before their Chief Executive Officer adopted a structural environmental policy focused on reducing CO_2 emissions (Interview British Petroleum, 2006).

However, with the development of codes, standards, policies and management systems, companies focused on the environment in a much more structural way. As Statoil explains (Interview Statoil, 2006), their environmental 'strategy gave all the environmental coordinators of each field a tool to understand what to work on; how to do things themselves' instead of relying on the few 'enthusiastic people' that drove environmental management before. In other words, these industry steering mechanisms can also serve as an internal instrument. Over the years, and especially with the development of environmental management systems, environmental management has become integrated throughout all layers of the companies. An environmental management system embeds environment as an issue within the company, also with the managers (Interview Wintershall Noordzee B.V., 2006). It also provides for internal standards that employees have to comply with (Interview Gaz de France, 2007). Finally, it provides for a system through which environmental regulation that comes from 'outside' can be implemented (Interview State Supervision of Mines, 2006).

Yet, the environmental policies, reports and management systems also have an external function. As mentioned before, the necessity of communicating the environmental management strategies of the offshore industry to the wider public was demonstrated during the Brent Spar incident. Certified environmental management systems are instrumental in that, because it show outsiders that the company has a system that is regularly audited (Interview Gaz de France, 2007). In addition, it allows governments to evaluate and control the environmental performance of companies (Karman & Tamis, 2000).

In the emergence of industry steering mechanisms, industry associations played a role as well. They can contribute to the development of industry steering mechanisms either by developing aggregate industry steering mechanisms or by supporting the development of corporate steering mechanisms by their members, i.e. the individual oil companies. An example of an aggregate industry steering mechanism developed in the offshore industry is the 2001 Sustainability Strategy of UKOOA. This strategy, among others, set targets and objectives for the environmental performance of the offshore industry operating on the UK continental shelf. However, UKOOA is the only industry association that developed an industry wide steering mechanisms, NOGEPA, OLF and the E & P Forum have not done so.

Another aggregate industry steering mechanism is a Norwegian one. Norsok standards is a Norwegian company that develops standards for among others the petroleum industry. Norsok standards developed an environmental care standard in 1994. According to them (Interview Standards Norway, 2006), this standard served two purposes. First, a cost reduction associated with having a single standard and solutions across the whole industry. Second, to facilitate the reduction of environmental impacts by simplifying existing national environmental regulations.

Another common practice is that industry associations support the development of environmental management within individual companies. This is often done through the publication of reports giving overviews of practices and technologies and the publication of specific guidelines. This practice is especially used by the E & P Forum, which has published reports on production water and developed guidelines on topics like waste management (1993 and 2008), environmental management (1997), environmental management systems (1994), flaring and venting (2000), etc.

With the emergence of industry steering mechanisms, a set of associated compliance mechanisms employed by the industry has emerged as well. One of these compliance mechanisms is environmental reporting. As noted above, environmental reporting has become a common practice in the offshore industry. However, not only individual companies have taken up environmental reporting, the industry associations do that as well. E & P Forum has published reports on the environmental performance of the offshore industry since 2001 and UKOOA and OLF since 1998. Nogepa has published such reports as part of the Environmental Covenant since 1995. This helps the companies to assess whether there is continuous improvement in their environmental performance.

Environmental management systems also have their own compliance mechanisms. Monitoring and review are an integral part of these systems. The actual discharges and emissions are measured and reported through an annual report and can also be used for the environmental report which are made public. Thus, overlapping in reporting exists. For example, reporting that was already done for the Environmental Covenant in the Netherlands has now become part of the environmental management system of Wintershall (interview Wintershall Noordzee B.V., 2006).

Another mechanism through which compliance is ensured is through auditing. Certified environmental management systems are subject to regular internal and external audits. These audits are used to assess whether the management system is implemented properly, whether internal objectives and policies are fulfilled, whether the company complies with legislative requirements and where improvements are possible (E & P Forum, 1994).

5.5.2 Implications for the national spheres of authority

The previous section outlines the self-governance within the offshore industry that developed during the 1990s. Interestingly, the national spheres of authority also emerged during that time. Industry steering and compliance mechanisms therefore influenced the national spheres of authority already during the emergence of the national spheres of authority itself. It is therefore not surprising that the national spheres of authority are partly characterized by industry involvement and joint initiatives. Industry self-regulation and national governmental regulation have become intermingled. This section will consider the intermingling of industry and national governmental steering mechanisms, which is visualized in Figure 6, in more detail. This will give a further understanding of how and why the national spheres of authority have emerged and what the authority of the industry is in these national spheres of authority.

The increased environmental awareness of the industry and the development of industry steering mechanisms has especially contributed to the development of respectively OGITF in the UK, Miljøsok and -forum, and the zero discharge approach in Norway and the Environmental Covenant in the Netherlands. These are all innovative steering mechanisms and/or deliberations structures which require an active attitude and participation of the industry, both in agreeing to objectives and strategies and in implementing these. The fact that the industry had already started to review its own environmental performance and ways to improve their performance has had a trade off in the joint initiatives developed in each country.

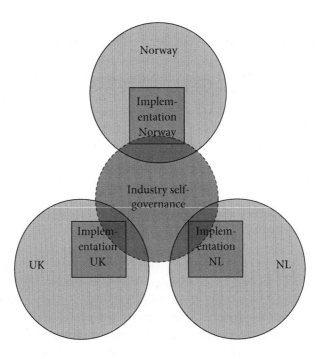

Figure 6. The national spheres of authority and industry self-governance.

In the first place, industry steering mechanisms have become intermingled with national steering mechanisms. On the one hand, national steering mechanisms have been developed that make use of the industry's willingness to consider environmental issues and regulations. For example, the Dutch Environmental Covenant originated from specific governmental policy, but has over the years developed into a shared initiative. The targets were initially set by the government, but there was some room to make these targets tailor fit for the offshore industry (Interviews FO-Industry, 2006; Institute for Marine Resources and Ecosystem Studies, 2006). After adoption of the targets by the government and the industry, the industry has been developing company environmental plans and has been taking measures to achieve these targets.

The Norwegian zero discharge approach has followed a similar path. The approach was initiated by the government, but was further operationalized by a working group in which the industry participated. When the objectives were agreed upon, the industry has developed individual plans to implement the zero discharge approach.

On the other hand, however, national steering mechanisms have influenced the steering mechanisms of the industry. This influence is exerted in three ways. First, because targets and objectives of national regulations are the basis for internal requirements set within companies (Interview Hydro, 2006). For example, Statoil has integrated the zero policy approach in their corporate environmental policy by adopting the zero harmful discharge objective internally (Interview Statoil, 2006).

Second, because national regulations contain procedures for developing implementation plans which shape the execution of corporate environmental policies. For example, the company environmental plans of the Dutch Environmental Covenant is not only an implementation document for the covenant itself, but is also used as an internal tool within certain companies (Interview Nederlandse Aardolie Maatschappij, 2005). The same goes for the zero discharge approach for which field specific plans had to be developed.

Finally, influence is exerted because national regulations push companies to develop environmental management systems. For example, the Environmental Covenant includes an objective that all companies have an environmental management system by 1995. As Gaz de France explains (Interview Gaz de France, 2007), this has given an impetus to companies to develop these systems. In the UK, one of the prerequisites for applying for a license is having an environmental management system (Interview Department of Trade and Industry, 2006c). The HSE regulation in Norway also require companies to develop environmental management systems (Interview Norwegian Pollution Control Authority, 2006a).

It should be noted that not only the steering mechanisms have become intermingled, but also one of the compliance mechanisms. Especially, the compliance mechanism of reporting often has a mutual purpose, i.e. one report serves both the internal environmental policy of the company and the national steering mechanism. For example, companies measure their own environmental performance (discharges and emissions) to assess whether there has been continuous improvement or whether their own objectives have been achieved. They and the industry associations often publish an environmental or sustainability report reflecting this assessment as well. In addition, national governments also require companies to report annually about their discharges and emissions.

In addition, because national regulations require companies to have an environmental management system, governments are also controlling the company environmental management system during inspections. For example, the SFT in Norway audits oil companies not only against national legislation, but also against the internal policies and management system of the company (Interview Norwegian Pollution Control Authority, 2006a). The Dutch state supervision of mines does so as well (Interview State Supervision of Mines, 2006). This means that what previously was an internal matter of a company, i.e. how they manage and control their environmental management system, has become a governmental issue as well.

However, the influence of industry self-governance onto the national spheres of authority is broader than only complementary and overlapping steering and compliance mechanism. The source of this broader influence is the deliberation structures between governmental and industry actors that have emerged. Examples are the Miljøsok and -forum and OGITF. However, also the Environmental Covenant provides a forum for deliberation. These forums provided regular interaction between the government and the industry, have resulted in mutual understanding between the government and the industry, and have led to consensus on principles or targets for future regulation. These initiatives would have been much less successful if the industry would not have had a cooperative attitude.

The broader influence of industry self-governance onto the national spheres of authority is demonstrated by a discourse dimension in the national spheres of authority that emphasizes the importance of stakeholder dialogue and industry responsibility. It also means that industry actors have gained a structural position in the actor dimension. And it has led to a power dimension that is not only characterized by regulator-complier relations but also by power that is shared between interdependent partners. In other words, the emergence of industry initiatives and regulations has not only influenced the steering mechanism and compliance dimensions of the national spheres of authority, but the other dimensions as well.

5.6 The renewed regional sphere of authority during the 1990s and 2000s

This section will shed light on the relation between the activities of the industry in environmental self-governance, the emerged national spheres of authority and the subsequent changes that have taken place in the regional sphere of authority as visualized in Figure 7.

There are three developments that have been instrumental in the emergence of national spheres of authority and industry self-governance. First, the development of national steering mechanisms, which expresses the approach each state takes towards eliminating pollution from the offshore industry. Second, the development of joint initiatives with the industry, which has changed the relationship between the governmental actors and the industry actors in the national spheres of authority. Third, there has been an increased participation of environmental NGOs.

To start with the latter, it was observed before that environmental NGOs were not part of the actor dimension of the regional sphere of authority. This changed in 1996, when Greenpeace presented their views on decommissioning before the SEBA meeting (Secretariat of the Oslo and Paris Commissions, 1996). It is highly likely that this change in the actor dimension was inspired by the Brent Spar affair of the year before. Since then, Greenpeace and World Wide Fund for Nature have attended SEBA meetings regularly (see summary records SEBA and OIC 1996-2009).

The 1992 OSPAR Convention allowed observers to be present at meetings of the OSPAR commission. However, meetings of committees and working groups were still closed to NGOs,

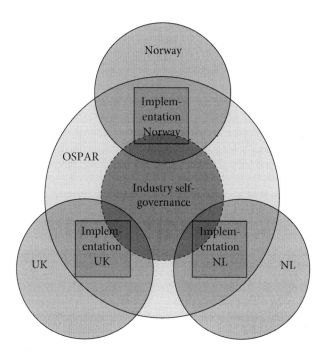

Figure 7. The regional and national spheres of authority and industry self-governance.

hence the practice of presentations at the beginning of these meetings. This changed in 1998, when OSPAR adopted new Rules of Procedures that allowed observers to other meetings as well (Secretariat of the Oslo and Paris Commissions, 2000). Since 1999, the E & P Forum, EOSCA, Greenpeace and World Wide Fund for Nature have therefore been present during the whole SEBA meeting. This change in the rules of the game also meant that NGO documents were taken up as an integral part of the documents for the SEBA meetings, instead of as a separate type of document.

Two other changes, unrelated to the emergence of national spheres of authority, deserve to be mentioned here as well. First, the E & P Forum changed its name to International Association of Oil & Gas Producers (OGP). Second, the OSPAR working group focusing on the offshore industry also changed name in 2000. The working group on Sea-Based Activities became the Offshore Industry Committee (OIC). With this change in name, the change in focus changed somewhat as well, i.e. the OIC no longer focused on dumping practices but solely on the offshore industry as a source of pollution.

The changed rules of the game and the direct participation of both industry associations in meetings where regional steering mechanisms are developed, is one aspect of the changed position of the industry. Another aspect is the active contribution of industry associations to the development of OSPAR decisions and recommendations. In the past, industry associations mostly reacted to plans of the GOP/SEBA or provided information about operational practices and technologies they used. However, the industry was more active in the development of the chemical regulations of OSPAR. The more direct and proactive involvement of the industry in the regional sphere of authority can be traced back to the industry's changed attitude towards the environment and the internal governing activities they have developed in the course of the 1990s.

For example, the Chemical Hazard Assessment and Risk Management model (CHARM), which is part of the Harmonized Mandatory Control Scheme for Offshore Chemicals, was a joint initiative. In 1996, the Paris Commission adopted the Harmonized Mandatory Control Scheme (HMCS) to reduce the discharge of hazardous chemicals. This scheme requires the offshore industry to submit applications for the use of chemicals. These chemicals have to be pre-screened to assess the hazardousness of the chemical. The industry's application is arranged through the Harmonised Offshore Chemical Notification Format. This information should enable the competent authority to permit the discharge of the chemical, to require substitution of the chemical, or to temporarily allow the use and discharge of the chemical when there are no alternatives. The information is also used to rank all chemicals so that less hazardous chemicals can be chosen. The CHARM model uses the internationally accepted Predicted Environmental Concentration : Predicted No Effect Concentration approach (Tatcher, Robson, Henriquez, Karman & Payne, 2005) to rank all chemicals. The CHARM model was developed under auspices of not only the OSPAR Parties, but also the offshore industry and the chemical suppliers (Tatcher *et al.*, 2005).

The changed attitude and involvement of the industry is also valued within the regional sphere of authority. The appreciation of active industry involvement is related to the national spheres of authority. National governmental actors participate in the SEBA/OIC meetings. Since their experiences with industry involvement in the national spheres of authority are positive, they are open for industry involvement in the regional sphere of authority as well. In addition, the developments in the position of the offshore industry and the joint initiative taken in developing the chemical regulations show that the discourses on industry responsibility and stakeholder dialogue have also affected the regional sphere of authority. It is important to note that there is a difference in the importance and impact of this discourse within the national and regional spheres of authority. Stakeholder dialogue was a key discourse that surrounded several initiatives for regular deliberation between governmental actors and industry actors initiated on the national level at the end of the 1990s. Within OSPAR, this discourse is not associated with a new initiative, but with including observers in the already existing committee focusing on the offshore industry, i.e. the Offshore Industry Committee.

Another change is that some elements of the regional steering mechanisms have been based on the risk based approach rather than on the precautionary approach. For example, the risk-based approach with its functional standards is used in the HMCS, where chemicals are ranked and substituted based on the risk they present to the marine environment. In addition, a 2001 decision on produced water sets a zero harmful discharge target for produced water, which has to be achieved by 2020. A consequence of taking a risk-based approach in regulating pollution from the offshore industry is that the risk based approach and the associated functional requirements (no harm) have become part of the discourse dimension of the regional sphere of authority.

There are two reasons – related to the industry self-governance – why functional requirements are possible. First, the national spheres of authority, through which these functional requirements need to be implemented, have deliberation structures and active involvement of the industry. Since functional requirements set the 'function' that should be reached but not how that should be reached, it is important to develop national or individual plans to achieve that function. National deliberation structures through which these plans can be developed together with the industry is vital for an effective implementation of functional requirements.

The second reason is that the industry is aware of its responsibilities and has developed internal governance mechanisms that make implementation of functional requirements possible. Consequently, when the industry is faced with functional requirements from OSPAR or through national steering mechanisms, they at least have internal mechanisms, e.g. environmental management systems, through which plans to achieve these targets in individual companies or at specific platforms can be developed and implemented.

Yet, it should be noted that the use of some functional requirements is complementary rather than substituting the traditional prescriptive requirements. Thus, the traditional precautionary approach with prescriptive requirements focusing on individual discharge streams is also still part of the discourse and steering mechanisms dimensions. This is for example confirmed by the Offshore Oil and Gas Industry Strategy. This is a long-term strategy adopted in 2003. The first long term strategies that serve as a guide for the work of OSPAR in each of its working area were adopted in 1998. The Offshore Strategy states that the precautionary approach is one of the principles adhered to and that environmental goals should be framed in 'measurable terms' as much as possible. More concrete examples of prescriptive standards are the 2000 decision on the discharge of organic phase drilling muds (muds based on animal, vegetable or mineral fluids) which simply prohibits the discharge of such muds or cuttings contaminated with these muds and the 30 ppm oil in produced water which has replaced the 40 ppm standards.

These changes have, of course, also affected the power dimension of the regional sphere of authority. Similar to what has happened in the national spheres of authority, the responsibility and authority have become more shared. The increased knowledge about discharges, emissions and possible harmful effects of the industry has empowered the industry in negotiations with governmental actors. For example, the Dutch State Supervision of Mines (Interview 2006) observes that the industry is much better informed and increasingly challenges the government to come with answers or explain unclear situations. Moreover, the fact that the industry has overcome its reluctance towards environmental issues and has become more proactive, the industry has become a more valuable partner in the development and implementation of environmental standards to minimize pollution from platforms.

Contrary to the national spheres of authority, however, the rules of the game still denote states, i.e. the Parties to the Convention, as the formal actors in both the development and the implementation of the regional spheres of authority. These states have voting power, while observers have not. In addition, the Parties to the Convention are responsible for implementing and complying with the agreed decisions and recommendations. The formal position of the states within the regional sphere of authority has therefore not changed.

The emerged national and industry approaches have also changed the national context in which implementation of regional steering mechanisms takes place. Implementation of regional steering mechanisms takes place in the UK, Norway and the Netherlands. With the emergence of national spheres of authority, the implementation of regional steering mechanisms has been influenced by certain characteristics of the discourse, actor, power and steering mechanisms dimensions of the national spheres of authority. This means that the regional and national spheres of authority partly overlap with each other (see also Figure 7).

The change in national context has been different for the UK than for Norway and the Netherlands. The national sphere of authority of the UK provides a better deliberation structure with the industry through OGITF and the two stakeholder forums in which implementation of regional steering mechanisms is discussed. At the same time, the implementation itself is no longer done through the combination between regulations and voluntary agreements with the industry, but through formal regulations alone. Thus while the deliberation of the implementation of OSPAR decisions and recommendations takes place in a more transparent manner with the industry through the stakeholder forums, DTI retains its power and responsibility the moment it develops formal regulations.

For Norway and the Netherlands, the implementation of regional steering mechanisms is no longer done through laws and regulations alone, but also through the zero discharge policy and the Environmental Covenant. Albeit both states already had their own ways to deliberate with the industry, i.e. mainly through working groups, this deliberation has intensified and has been formalized through the zero discharge policy and the Environmental Covenant, also for the implementation of regional steering mechanisms. The trend has been to make the industry more directly responsible for developing plans to achieve targets and limits set at the regional level. Moreover, it also allows the industry to link up long-term national strategies (i.e. zero discharge policy and the Environmental Covenant) and the requirements set at the regional level with internal industry policies (environmental codes and management systems). Industry self-governing initiatives are therefore often developed in line with regional and national steering mechanisms so that a comprehensive set of standards can be implemented through the environmental management system.

The compliance dimension has not changed. The main regional compliance mechanism is still reporting. For example, reporting of annual discharges still continues. One compliance mechanism that has improved is the reporting of the implementation activities of each Party to the Convention. This improvement was initiated in 1994 when it was decided to develop guidelines for implementation reports. This should make the implementation reporting procedures more transparent and easy to follow (Secretariat of the Oslo and Paris Commissions, 1997). In order to do so each OSPAR decision and recommendation should include dates for when the implementation reports should be submitted. Based on these reports, implementation assessments of a decision or recommendation are made and then discussed within SEBA/OIC meetings.

The main national compliance mechanisms have also remained unchanged. Each country relies on a combination between inspections and reporting. The inspections are important because they are related to the permit system that exists in especially Norway and the UK and that are used to implement OSPAR requirements. Reporting has been an important aspect of the zero discharge policy of Norway and the Environmental Covenant in the Netherlands as well.

Finally, the new discourses on stakeholder dialogue and industry responsibility and the joint zero discharge approach and Environmental Covenant have changed the power relations with regard to implementation practices in the regional sphere of authority. Power with regard to implementation and compliance has become much more shared between the government and the industry.

5.7 The authority of the state during the 1990s and 2000s

After the developments of the 1990s and early 2000s, it is time to assess changes in the authority of the state. The changes in authority stem from two underlying trends. The first is the increase in spheres of authority that develop steering mechanisms. Besides OSPAR, there are the national governments and the industry itself that have developed spheres of authority to develop environmental regulation over the last decade. The authority of the state (and of the industry) therefore no longer only depends on its authority within OSPAR, but also its authority nationally and with regard to the industry steering mechanisms.

Second, while the national spheres of authority shows an important role for the state, the industry gained its own place within the national spheres of authority and developed a system of self-governance. The increased involvement of the industry in the environmental governance of offshore oil and gas production might have led to a loss of authority for the state.

5.7.1 Authority in developing steering mechanisms

Traditionally, the development of steering mechanisms to govern the environmental impact of the offshore industry was mostly in hands of the state. Decisions on standards and requirements were made within the GOP/SEBA meetings and the Paris Commission. The NSMCs facilitated these decisions, because they provided the North Sea states with a forum for consensus building. Yet, the industry had authority as well, based on sharing their views and knowledge about environmental effects and existing practices and technologies to deal with these effects. As we have seen, there were several moments in the decision making process in which the industry could do that, i.e. by lobbying their government, by giving presentations at the beginning of GOP/SEBA meetings and during meetings specifically organized to share knowledge and ideas between the GOP and the industry.

The authority of the state in developing steering mechanisms has gained a national dimension through the emergence of national spheres of authority. The governmental actors in the UK, Norway and the Netherlands have expanded their influence on steering mechanisms by developing national steering mechanisms. However, how they have done that differs. The UK moved away from agreements with the industry to developing formal regulations. The authority to do so was granted to DTI by formal law in 1999. In a way, they have become less dependent on the industry for governing the environmental impact of the offshore industry, because the real need to have the industry's agreement has become less (Interview British Petroleum, 2006). On the other hand, active input and support from the industry has been sought through OGITF and the two deliberation forums.

In contrast, in Norway and the Netherlands, the active participation of the industry is not only arranged through deliberation forums, but also by developing joint initiatives with industry participation from the start. Both countries have allowed the industry to join in the debate about which environmental issues are important and which targets can be set. Moreover, certain responsibilities have been put into the hands of the industry by requiring the development of company or field specific plans. In doing so, the authority for developing steering mechanisms is partly put into the hands of the industry, allowing the industry to be flexible in how it deals with issues. Of course, boundaries and specific moments are built in to check the work of the industry, i.e. industry plans need to be approved and formal regulations setting minimum limits and requirements are in place in both countries.

The result, in terms of authority of the state, seems to be mixed. On the one hand, with a national sphere of authority, national governments have taken matters into their own hand by developing national steering mechanisms. This has increased the authority of the state, because the national governments have been able to develop steering mechanisms that go beyond OSPAR. Yet, the authority of the industry has increased as well. They are actively involved in decision making and their participation is valued. What is more, it is a corner stone of the approaches developed in the Netherlands and Norway.

Moreover, the industry itself has been active in developing steering mechanisms. Most of these are company specific, which has increased the authority of individual companies over their own environmental performance. Instead of trying to influence governmental based steering mechanisms, the industry have increased their authority over how to ensure a sufficient level of environmental performance by developing company specific policies. In some rare cases, i.e. in the case of the sustainability strategy of UKOOA and the Norsok standards, the industry has developed a steering mechanism on the industry level.

Although the development of self-governance has been an industry matter, the state has some authority with regard to the requirements that the companies set for themselves. The state has this influence, because OSPAR and national requirements are the basis for the development of company environmental policies (Interview Hydro, 2006). This authority is even stronger in Norway and the Netherlands, where company environmental plans are integrated with national steering mechanisms, i.e. the joint initiatives of the zero discharge policy and the Environmental Covenant. What is more, the government has to approve the plans that are made by the companies. Finally, the authority of the government also lies in ensuring that all companies have developed an environmental management system that integrates the environment in all levels of the company. However, it has to be said that the governmental requirement of having an environmental management system is a reaction of a trend that already took place within industry.

All in all, the strong link between the national spheres of authority, OSPAR standards and industry steering mechanisms has not only meant an increase in authority of the industry over how to control their own environmental performance, but also some authority of national governments over the industry steering mechanisms.

The increased authority of the industry in the environmental governance of offshore oil and gas production has affected the OSPAR sphere of authority as well. The changes in the OSPAR sphere of authority are caused by the difference in attitude of the industry and their expertise with regard to environmental policies. In line with this development, the most important OSPAR meeting in which decisions are taken is since 2000 open to industry associations and environmental NGOs. Some argue that for the industry this has not made a difference, since the industry always made sure that national delegations were very well briefed (Interview OSPAR Secretariat, 2006) Moreover, the presence of the industry serves a very practical purpose, i.e. they have the expertise about operations and technologies available (Interviews International Association of Oil & Gas producers, 2006; Norwegian Pollution Control Authority, 2006a). In that sense, the authority of the industry still seems to be based on providing information and sharing their expertise. And whether this is done through national delegations or directly at the table during the negotiations does not have to matter.

Yet, there is more to it than that, because the industry has become better in challenging the government. For example, the industry has become better in specifying why stricter environmental regulations are not desirable (Interview Norwegian Pollution Control Authority, 2006a). The industry has also become better in pointing out the weaknesses and vagueness in arguments used or plans made by the government (Interview State Supervision of Mines, 2006). The changed attitude of the industry, the development of self-governance in the industry and the recognition that industry input is important in developing regulations were instrumental in this empowerment of the industry. In other words, the industry has increased its authority within OSPAR.

Besides increased participation of the industry, environmental NGOs have started to participate within OSPAR meetings for the first time since the Paris Convention was established as well. Before the Brent Spar incident in 1995, environmental NGOs hardly focused on the offshore industry as a source of pollution. Brent Spar served as an eye opener for these NGOs. They therefore started to present at SEBA meetings and participate in OIC meetings.

To sum up, the changing authority in the environmental governance of offshore oil and gas production in the end of the 1990s and early 2000s was characterized by several trends. First, the governments have increased their authority over the development of steering mechanisms through the national spheres of authority and because of their authority over the development of industry steering mechanisms. Second, the industry has done so as well. This has not necessarily been at the expense of governmental authority, rather authority has become increasingly shared between governmental and industry actors. However, this does mean that the authority of the governmental actors has seen a relative decline. After all, the authority in developing steering mechanisms is no longer based on OSPAR alone, but also on a national sphere of authority with joint initiatives and industry steering mechanisms. Third, environmental NGOs have gained some authority, although this is very limited compared to the governmental and industry actors involved in the environmental governance of offshore oil and gas production.

5.7.2 Authority in generating compliance

The growth in the sources of authority that develop steering mechanisms to govern pollution of the offshore industry is accompanied with a growth in sources that develop compliance mechanisms as well. Traditionally, the regional sphere of authority required its members, i.e. states, to report discharges and failures to meet the 40 ppm oil in produced water target. The results of these reports were combined in annual and triennial reports, which were subject to review in the meetings of the GOP/SEBA and the Paris/OSPAR Commission.

The states, i.e. the UK, Norway and the Netherlands were each responsible for the implementation of and ensuring compliance with regional steering mechanisms. In practice, that means that these states required companies operating on their continental shelves to report their discharges. Besides that, these states conducted inspections, although the effectiveness of the earlier inspections can be put into question because a 40 ppm target was not based on ad hoc basis, but on sampling. Finally, all three states also monitored the marine environment, often with the help of the industry. This means that the states were depended on the industry for information concerning compliance.

After information gathering took place nationally, the states reviewed itself and each other on the regional level.

Even though national spheres of authority emerged from the mid-1990s onwards, the way in which compliance is ensured did not change much. The main change has been that reporting has become more important, because reporting is the main compliance mechanism in the joint initiatives of the Norwegian zero discharge policy and the Dutch Environmental Covenant. This does mean, however, that the dependence on the industry for compliance related information has grown. The industry is the one that has to provide reliable information on discharges and emissions. Since, reporting has become a more important compliance mechanisms, the authority of the industry has become more important as well. Still, the state is the one that uses the information to assess compliance and can use it to intervene when necessary.

However, the emergence of industry steering mechanisms has been more important in changing the way in which the state ensures compliance. With the emergence of industry steering mechanisms, reporting and auditing emerged as industry compliance mechanisms as well. The reporting of the industry serves three purposes: to fulfil the reporting requirements of the government, to assess environmental performance internally and to publish an environmental report to a wider public. This does not have an impact on the authority of the state in ensuring compliance, because the state still sets requirements to the information gathered which the industry has to meet. However, it does mean that the efforts to ensure compliance on the part of the industry are growing.

What is more relevant for the authority of the state is that the industry has developed internal steering mechanisms that are audited. The consequence is twofold. On the one hand, the industry also audits its own environmental policies and environmental management systems. Moreover, if a company has a certified environmental management system, it is audited externally, by the certification body. According to Statoil (Interview Statoil, 2006), these external audits of certification bodies lessens the need for the government to audit the environmental management system. This means that the industry itself has taken the responsibility of ensuring compliance more seriously. The industry has developed ways to check its own compliance, which are separately from the compliance mechanisms used by the governmental actors.

On the other hand, the development of internal environmental policies and especially the environmental management systems has made inspections focused on environmental performance by governmental actors easier. Governmental inspectors can audit the environmental management system and the way in which regional and national steering mechanisms are integrated into that system. They are allowed to do so, because the requirement of having an environmental management system is laid down by OSPAR and by the individual states. These inspections allow a better judgement of compliance and ways to improve compliance. In other words, the authority of the state in ensuring compliance has improved because of the existence of industry steering mechanisms.

All in all, the industry has increased its efforts in generating compliance. Reporting has become more important for them, while they also arrange audits to certify their management system. At the same time, the state has somewhat more grip on ensuring compliance, because the reporting of the industry has become more reliable and because the inspections of the environmental management system as a whole. In other words, more efforts are done to ensure compliance, but these are not only done by the state, but by the industry as well. This means that the authority of the industry over ensuring compliance has increased, with the result that the authority of the state in ensuring compliance has seen a relative decline.

5.8 The emergence of an European Union sphere of authority during the 2000s

Partly parallel to and partly after the emergence of national spheres of authority and industry self-governance, the EU started to emerge as a regulatory force in the environmental governance of the offshore industry as well. Since 1997, several existing European Union steering mechanisms have become applicable in the Exclusive Economic Zone and thus also to the whole offshore industry operating in the UK and the Netherlands (Norway is not a member of the EU). In addition, new directives and regulations have been adopted that apply to the offshore industry as well. This has meant that the EU also became a sphere of authority for the offshore industry. Unfortunately, the scope of this research does not allow for a detailed analysis of all the dimensions of the EU sphere of authority. The focus will therefore lay on the actor and steering mechanisms dimensions.

The first example of an EU steering mechanism is the Environmental Impact Assessment Directive. This Directive was already agreed upon in 1985, but it was not until the amendments of 1997 that the Directive also applied to the 'extraction of petroleum and natural gas for commercial purposes'. Only extraction activities of 500 tonnes/day for petroleum and 500 000 m³/day for gas or more need to be preceded by the Environmental Impact Assessment.

In addition, in 2001, the EU adopted the Strategic Environmental Assessment Directive. The purpose of this Directive is to ensure that environmental consequences of certain plans and programmes are identified and assessed during their preparation and before their adoption. The difference between this Directive and the Environmental Impact Assessment Directive is that a Strategic Environmental Assessment is done to evaluate a whole plan or programme, while an Environmental Impact Assessment evaluates a specific project. This means that a Strategic Environmental Assessment is undertaken much earlier in the decision-making process than an Environmental Impact Assessment.

The Birds and Habitats Directive are part of the nature conservation policy of the European Union. When transposing these Directives into national law, the UK and Dutch governments decided not to implement this directive outside their territorial waters. This decision of the UK government was challenged by Greenpeace in a court case in 1999 (Rowan-Robinson, 2000). The ruling of the judge was in favour of Greenpeace, which basically meant that the Habitats Directive should also be applied to the whole UK Continental Shelf (Interview Department of Trade and Industry, 2006a). In 2004, the EU decided that the Birds and Habitats Directives should be applied to the marine environment by 2008 (European Commission, 2007).

Another example of such an EU directive is the Integrated Pollution Prevention and Control Directive. This Directive sets out standards to control the pollution from industrial installations in an integrated way, i.e. a permit covering all pollution is required for bigger industrial installations. Since the end of 1999, this permit is required for new installations and existing installation that are subject to a substantive change. All installations both on land and at sea need to have a permit by the end of 2007. The most important aspect of this permit is the requirement to have Best Available Techniques. What the best available techniques are is arranged through a process in which the government and industry develop a reference document for that specific sector, the so-called BREF.

A directive focusing on air emissions is the 2003 Directive on a European Union Emissions Trading Scheme. This Directive provided for the establishment of an Emissions Trading Scheme for CO_2 between companies within the EU as of 2005. The EU trading scheme preceded the emissions trading scheme which has been set up worldwide under the Kyoto Protocol in 2008. Only companies with a combined power of 20 mw or more fall under the emission trading scheme. In practice, this means that all operators operating on the (Dutch) continental shelf have to participate in the European CO_2 emission trading scheme (FO-Industrie, 2008).

The most recent piece of European legislation that affects the offshore industry is the Regulation for Registration, Evaluation, Authorisation and Restriction of Chemicals (REACH) which entered into force in June 2007. This regulation requires that all chemicals are subject to registration and evaluation, while the most hazardous chemicals are subject to authorisation and restrictions. In other words, just like OSPAR, the EU has turned its attention to chemical pollution of the offshore industry.

All in all, the development of European legislation becoming applicable to the offshore industry meant that the offshore industry has been confronted with a new set of steering mechanisms. Furthermore, these steering mechanisms co-exist next to the OSPAR, national and industry ones, which differ in the scope they have. While the latter develop steering mechanisms which are explicitly meant for the offshore industry, the EU directives have a much broader scope, i.e. they apply to other industrial sectors as well. In addition, while the OSPAR, national and industry steering mechanisms are focused on reducing discharges to water, the EU directives cover air emissions, environmental assessments and the protection of marine flora and fauna.

Besides a new set of steering mechanisms, new actors have become involved in the environmental governance of offshore oil and gas production as well. These actors are involved because they are part of the actor dimension of the European sphere of authority. Most relevant for the environmental governance of the offshore oil and gas industry is the fact that the offshore industry and the regulatory actors of the offshore industry are much less involved in the EU sphere of authority. Those people that are directly involved in OSPAR are also involved in the national regulation of the offshore industry (Interviews British Petroleum, 2006; FO-Industry, 2006), but not always within the EU.

This is caused by the fact that the scope of the EU steering mechanisms is much broader than the offshore industry alone. For example, the Ministers of Environment of different member states are much more involved than the Ministers that are responsible for the offshore industry. Similarly, the offshore industry did not have close relations with the European institutions either, because the EU was not relevant to the offshore industry.

As a reaction to this development the industry and governmental actors have been putting much effort in building these relations. Yet, both have experienced difficulties in influencing the development of EU steering mechanisms early enough in the process to make a difference (Interview Consultant offshore industry, 2006). For example, DTI explained that it can also be difficult to ensure that EU legislation is appropriate for the offshore industry, because it is sometimes difficult to make clear what the differences are between land-based industries and the offshore industry (Interview Department of Trade and Industry, 2006c). All in all, this results in questions concerning the extent in which the government of the UK and the Netherlands are able to influence EU legislation to make them more appropriate for the offshore industry (Interviews

British Petroleum, 2006; Consultant offshore industry, 2006; Department of Trade and Industry, 2006c).

Together with the EU steering mechanisms, an new compliance mechanism has also become part of the environmental governance of offshore oil and gas production, namely the infringement procedure. The infringement procedure is a procedure that the European Commission can start when a member state does not implement legislation as it should. Since this procedure has the possibility to end with a court case in front of the European Court of Justice, the compliance mechanism is a powerful one.

5.9 Renewed national and regional spheres of authority during the 2000s

The emergence of the EU sphere of authority brings about the question whether the national and regional spheres of authority have renewed itself as a result of this policy change. This section will first discuss the renewal of the national spheres of authority. This analysis will be limited to the UK and the Netherlands for two reasons. First, since Norway is no EU member state, it is not involved in the EU sphere of authority, while the UK and the Netherlands are. Second, even though Norway implements EU environmental legislation within the Agreement on the European Economic Area, none of the interviewees have mentioned the implementation or relevance of EU legislation for the offshore industry. This means that it is assumed that the Norwegian national sphere of authority has not renewed itself as a consequent of the emergence of the EU sphere of authority. After dealing with the national spheres of authority, the renewal of the regional sphere of authority will be discussed.

5.9.1 Renewed national spheres of authority

The co-existence and the interaction between the UK and Dutch spheres of authority and the EU sphere of authority are visualized in Figure 8.

Yet, the relationship between European steering mechanisms and the national ones differs per EU member state. One element that differs is the extent to which and the way in which the UK and Netherlands are implementing these directives for the offshore industry. For example, the habitats directive started to play a role in the appointment of new areas for exploration and production in the UK in 1999 (Rowan-Robinson, 2000). What is more, the Offshore Petroleum Activities (Conservation of Habitats) Regulations 2001 were developed for that purpose. In the Netherlands, the Nature Conservation Laws have not been amended to apply the birds and habitats directives offshore yet (FO-Industrie, 2008). In addition, the designation of special areas of conservation or special protected areas are still not formalized in both countries (FO-Industrie, 2008; Joint Nature Conservation Committee, 2009).

Another example is the Environmental Impact Assessment Directive, which is in the Netherlands formally implemented through the Environmental Protection Act. In practice, the development of impact assessments for offshore oil and gas production is guided by a sector-wide environmental impact assessment. This generic environmental impact assessment was initiated and developed by NOGEPA and serves as a document which can be used as a basis for the development of site-specific assessments (Van Oosterom, 2001).The UK already worked with

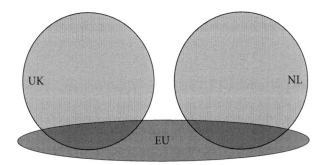

Figure 8. Co-existence of the UK, Dutch and EU spheres of authority.

impact assessments in granting licenses for areas within 25 miles of the coast (Rowan-Robinson, 2000). Since the 1999 Offshore Petroleum Production and Pipelines (Assessment of Environmental Effects) Regulations, all production activities that cross a certain threshold have to submit an environmental statement, which is based on the impact assessment, when seeking consent for drilling or field development. On top of that, the UK is working with strategic environmental assessments since 1999, thus preceding the EU Strategic Environmental Assessment directive of 2001. In contrast to the UK, the Netherlands has not undertaken a strategic environmental assessment for the offshore industry (yet).

With regard to the IPPC directive, the same differences prevail. The UK has developed regulations to implement this directive for its offshore installations larger than 50 mw through the 2001 Offshore Combustion Installations (Prevention and Control of Pollution) Regulations. In contrast, in the Netherlands the interpretation is that the Dutch offshore production installations do not fall under the directive (FO-Industrie, 2006, 2008).

Still, generally speaking, during the last decade the EU has slowly appeared as a regulating force for the offshore industry. This has, in the first place, led to a change in the steering mechanism dimensions of the British and Dutch spheres of authority. As indicated, there are slight differences in how European legislation has exactly affected the steering mechanism dimension in the Netherlands and the UK as a result of differences in interpretation and implementation. Yet, the impact is apparent in both member states.

However, there is also an important aspect to this change that the UK and the Netherlands share. The British sphere of authority was characterized by some formal regulations, a Pilot trading scheme for flaring and the industry sustainability strategy as the main steering mechanisms. However, the implementation of EU directives have resulted in a whole set of new – formal – regulations. Currently, the amount of formal regulations put the joint and industry initiatives in the shadow.

Similarly, the Dutch sphere of authority depended on the Mining Laws together with the Environmental Covenant. Now that many of the issues, especially air emission issues, which were dealt with within the covenant have become subject to EU legislation, the focus has shifted to formal laws and regulations. The emphasis is therefore shifting from cooperation with the industry to a more regulatory approach in which companies have to meet norms. In other words, in both the UK and the Netherlands there has been a shift away from joint initiatives with the industry to command-and-control through formal, governmental regulations. Yet, in the UK this shift has gone further than in the Netherlands, because the renewed UK sphere of authority consists of only formal regulations, while the Netherlands still has the joint initiative of the Environmental Covenant as an important source of steering.

Another change in the steering mechanism dimensions, which the Netherlands and the UK share, is the increased need to coordinate the great number of steering mechanism that are now part of the environmental governance of the offshore industry. The Netherlands initially had to deal with the Mining Laws, the Environmental Covenant and OSPAR decision and recommendations. The first two were used to implement the OSPAR regulations. Moreover, the deliberation structure within the Environmental Covenant was very important in integrating and implementing OSPAR requirements into the Dutch sphere of authority. However, because of this policy change, the EU directives and regulations also need to be integrated and implemented within the sphere of authority. This has become a complex issue, because the European directives have a broader scope than the offshore industry. Therefore they also fall under other regulatory authorities, i.e. nature conservation falls under the Ministry of Agriculture, Nature and Food Quality, while the environmental assessments and air pollution are part of the Ministry of Spatial Planning, Housing and the Environment's portfolio. Coordination does therefore not only take place between the Ministry of Economic Affairs and the offshore industry in the framework of the Environmental Covenant, but also has to be sought with the other two Ministries and the laws they use to implement EU legislation (such as the Environmental Protection Act and the Nature Conservation Laws).

In the UK, the situation has been different. DTI has taken up the issue of coordination by developing sector specific regulations to implement EU regulations. This is a different approach than the Netherlands, where the main laws covering that issue are amended and implementation in the offshore industry has, in the end, to be arranged through the company environmental plans within the Environmental Covenant. In the UK, the regulations, which all require a permit, ensure appropriate implementation of EU legislation for the offshore industry. The subsequent plethora of permits that has arisen is one of the side effects of all the regulations that DTI has developed in the past decade to implement EU legislation; the offshore industry needs a permit to discharge oil, a permit for all use and discharge of offshore chemicals, an 'integrated' permit for combustion plants and a CO_2 permit for the emissions trading scheme. While the intention was to develop an integrated permit (Interview United Kingdom Offshore Operators Association, 2006), only the application of these single permits is covered by an integrated application process (Interview Department of Trade and Industry, 2006b). In a sense, this also presents coordination problems for DTI and the offshore industry.

The changes in the steering mechanisms are related to some small changes in the actor dimension. In principle, the type of actors are still the same as in the initial national spheres of authority, i.e. the actor dimensions are still characterized by the regulatory governmental agencies for the offshore industry and the environmental authorities, the industry associations and the individual oil companies. Yet, in the Netherlands, the Ministry of Environment has become more involved, because that Ministry is responsible for the implementation of EU environmental legislation. For the UK, the increased authority of DTI in developing regulations for the offshore industry is an important change. Dti gained the power to develop secondary regulations in 1999. This change in the rules of the game dimension of the UK sphere of authority is not specifically related to the emergence of the EU as a source of authority. This change is, however, instrumental in the way in which the UK sphere of authority changed as a result of the emergence of EU steering mechanisms.

The developments in the steering mechanism dimensions have also caused a change in the discourse dimensions of the British and Dutch spheres of authority. The discourse of governing through dialogue with the industry has been pushed to the background by a discourse about how heavily regulated the industry has become. For example, as someone from DTI (2006c) noted 'in 10 years time, the UK moved from a situation in which there was relatively little environmental control to a situation in which the offshore industry is on a par with other industries and even ahead of some industries'. Someone from the industry (Interview British Petroleum, 2006) mentioned that 'the UK has tended to move away from those voluntary bilateral agreements and has moved towards using the powers to apply legislation and it is tended to apply more and more legislation'. In the Netherlands, this discourse is also expressed. For example the Dutch industry argues that they are quite tightly embedded in law and regulations (Interview Nederlandse Aardolie Maatschappij, 2005) and that all new regulations have been imposed while self-initiative from the industry is waning (Interview Wintershall Noordzee B.V., 2006).

In addition, the fact that the new steering mechanisms come from an unfamiliar source of authority also affects the discourse dimension. This is in the first place expressed through a discourse about the distance the offshore industry and their regulators perceive with the EU. This distance is bigger than with OSPAR, because those people that are directly involved in OSPAR are also involved in the national regulation of the offshore industry (Interviews British Petroleum, 2006; FO-Industry, 2006). This distance and the complexity of EU decision making processes makes that the extent in which the government is able to influence EU legislation to make them better appropriate for the offshore industry is questioned (Interviews British Petroleum, 2006; Consultant offshore industry, 2006; Department of Trade and Industry, 2006c).

Another impact of the EU sphere of authority is a new discourse in which the main driver for environmental regulations for the offshore industry in recent years is believed to be the EU and not OSPAR. As Rowan-Robinson (2000, p. 281-282) observes 'where legislation has been introduced over the intervening years, this had tended to be a response to commitments under international conventions and European legislation'. This was also confirmed by several interviewees (Interviews Consultant offshore industry, 2006; Department of Trade and Industry, 2006b; FO-Industry, 2006; Netherlands Oil and Gas Exploration and Production Association, 2006)}. For example, one argues that everything that is new comes from the EU (Interview Netherlands Oil and Gas Exploration and Production Association, 2006) while another mentioned that EU issues might overrule OSPAR in the future (Interview Nederlandse Aardolie Maatschappij, 2005)

Finally, the increase in formal regulations stemming from the EU have led to a discourse within the offshore industry around the heavy administrative load and the lack of added value it will have in preventing pollution from the offshore industry. For example, especially the added value of the chemical regulations (REACH), the ippc directive and the CO_2 emission trading scheme is questioned (Interviews Gaz de France, 2007; Wintershall Noordzee B.V., 2006). UKOOA (Interview 2006) is even willing to go further than that: 'what it actually has done is having a negative effect, not on environmental performance in the strict sense, but on the willingness within the operators to go beyond regulations and make continual improvements'. The industry feels it is overregulated. While it was clear that oil based muds and hazardous chemicals were a threat to the environment, it is the question whether further measures will be cost-effective (Interview Nederlandse Aardolie Maatschappij, 2005). These observations in turn enhance the discourse that questions the value of more and more (EU) regulations.

The emergence of an EU sphere of authority and some of the changes in the discourse dimensions have affected the power dimensions of the UK and the Dutch spheres of authority as well. Even though the regulatory state actors, i.e. DTI for the UK and the Ministry of Economic Affairs in the Netherlands, are in some sense confronted themselves by EU legislation, they become one of the responsible parties when this legislation needs to be implemented. Since EU legislation is mostly implemented by formal, national legislation, the government-industry relations shift back again to a command-and-control relationship, instead of the relationship of shared responsibility which has characterized the national spheres of authority since the end of the 1990s. The decreasing relevance of joint and industry initiatives in the environmental governance of offshore oil and gas production only adds to this change.

It should be noted, however, that the partner relationship between the government and the industry still exists when it comes to deliberating how EU regulations affect the offshore industry and how they can be integrated in the already existing initiatives and plans. An important difference between the UK and the Netherlands in this regard is that in the Netherlands the challenge of implementing EU legislation is dealt with within the deliberation structure of the Environmental Covenant. In contrast, the UK uses a permit system. This gives the government-industry relations in the UK a more hierarchical touch than in the Netherlands. In the Netherlands, the government has taken the effort to provide an overview of relevant EU legislation in its guidelines for the new company environmental plans in 2006 and the relationship between EU legislation and the existing targets of the Environmental Covenant (FO-Industrie, 2006; Interview FO-Industry, 2006). The industry subsequently integrated the EU requirements in their company environmental plans. This has all taken place within the structure of meetings, committees and deliberation that already existed and where both the government and the industry had already gotten used to.

This difference between the UK and the Netherlands also comes back in the compliance dimension. In the UK, the permit system has as a consequence that inspections have become a much more important compliance mechanism, at least for ensuring compliance with EU legislation. Because of the comprehensive nature of the Dutch Environmental Covenant and subsequent integration of EU legislation in this framework, reporting has become the main compliance mechanism for EU legislation in the Netherlands. Reporting procedures under the Environmental Covenant has thus remained the key compliance mechanism for all steering mechanisms in the Dutch sphere of authority. This also adds to the less hierarchical structure in which the government and industry have to cooperate in the Netherlands than government-industry relations existing in the UK.

Another change in the compliance mechanism is that with the emergence of EU legislation, the EU infringement procedure can also start to play a role in the environmental governance of the offshore industry. In cases where EU member states have not properly implemented the EU directives in their national laws, the European Commission can start the infringement procedure. However, it is not expected that this compliance mechanism will play a big role in the environmental governance of the offshore industry, because the offshore industry will only be an indirect target of such a procedure. This is related to the nature of EU legislation, which is broader than the offshore industry alone. Thus, even though the infringement procedure targets a member state, most of the time, other state actors than DTI and Ministry of Economic Affairs, i.e. the Environmental Ministries of both countries, are the ones that have to respond to the infringement procedure in first instance.

5.9.2 Renewed regional sphere of authority

The influence of the EU sphere of authority on the regional sphere of authority is much more indirect than on the national spheres of authority. This is due to the fact that EU member states, like the UK and the Netherlands, are involved in the development and implementation of EU legislation and not the OSPAR Commission or bodies. In addition, the relationship between the EU and the OSPAR bodies is more competitive and complementary in nature, since the scope and members of both partly overlap, i.e. European Seas versus the North-east Atlantic Ocean. Similar to the national spheres of authority, the influence of the EU as a regulating force for the marine environment is broader than the offshore industry alone. Taking a closer look at the EU steering mechanisms developed in the last 10 years that focus on the marine environment, it can be concluded that they not only affect the work on the offshore industry, but also other issue areas of OSPAR, such as eutrophication and marine protected areas.

For the offshore industry, EU steering mechanisms both overlap and complement those of OSPAR. Especially the recent EU regulation on chemicals is similar to the already existing Harmonized Mandatory Control System for the Use and Reduction of the Discharge of Offshore Chemicals of OSPAR. In issues like air emissions EU legislation does complement OSPAR's work. Where OSPAR has always focused on discharges of the offshore industry to the sea, the EU legislation that has become applicable to the offshore industry is much more focused on air emissions; for example, through the ippc directive and the CO_2 emissions trading scheme.

Regardless of the overlapping or complementary nature of EU steering mechanisms, a discourse around the need for coordination has developed within OSPAR. This need was already discussed in various committees and meetings within OSPAR in 1998 (Secretariat of the OSPAR Commission, 1998). An information document, which lists subjects of interest and possible ways to cooperate with the European Community in the future, was the result. For the offshore industry, the issues of hazardous substances and radioactive substances were mentioned as such a subject of interest (Secretariat of the OSPAR Commission, 2001).

With the adoption of the EU REACH regulation in 2006, the coordination of chemical policies of the EU and OSPAR became a necessity. Moreover, this necessity has become a new discourse. This coordination was put on the agenda by EOSCA in 2005 (Secretariat of the OSPAR Commission, 2005). By 2008, a more detailed analysis into the implications of REACH for the HMCS of OSPAR was done (see Annex 4 of Secretariat of the OSPAR Commission, 2008). The conclusion was that REACH applies not only to territorial waters, but also to the exclusive economic zones and the continental shelf. Furthermore, REACH applies to all types of chemicals. This means that REACH goes further than the HMCS which allows for certain exemptions. On the other hand, the HMCS goes further in requiring hazardous chemicals to be substituted by less hazardous chemicals. Despite these conclusions, real coordination to harmonize the HSMC with REACH is limited. In addition, contact with the European Chemicals Agency has not been established yet (Secretariat of the OSPAR Commission, 2007, 2008).

Another important implication of the policy change around the EU becoming a regulating force is the possible redefinition of the role that both the EU and OSPAR have in protecting the marine environment and in regulating the offshore industry. The application of the Birds and Habitats directives onto the exclusive economic zones and continental shelf were a first step in this regard. The EU is currently also developing an integrated maritime policy with a marine strategy to protect the marine environment of the European seas. This overlaps with work of OSPAR on biological diversity and ecosystems. One of the ideas that exists is that OSPAR could become a regional executer of the integrate EU maritime policy (Interview Ministry of Transport, Public Works and Water Management, 2006).

At the moment, however, both OSPAR and the EU co-exist next to each other (Figure 9). Both develop their own steering mechanisms in a sphere of authority that already existed before that. OSPAR is seeking coordination and information sharing with the EU but not more than that. This also means that the interaction between actors and the rules of the game dimension, the power dimension and the actor dimension within the regional sphere of authority central in this chapter (i.e. OSPAR) has not been subject to change, but that the changes have been limited to coordination and overlap between the steering mechanism dimensions and one change in the discourse dimension.

Even though the emergence of the EU sphere of authority has not led to a renewal of the OSPAR sphere of authority, there have been two changes in the regional sphere of authority that

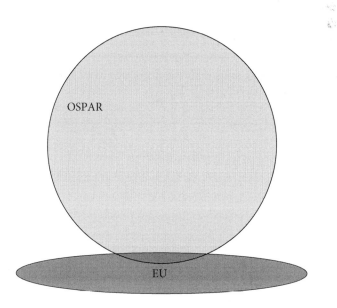

Figure 9. The co-existence of the regional and EU sphere of authority.

should be mentioned. First, the initial regional sphere of authority was marked by the impetus that the North Sea Ministerial Conferences gave to the paris/OSPAR Commission. In principle, the policy changes of emerging national spheres of authority, industry self-governance and EU steering mechanisms have not affected the NSMC - OSPAR relationship. The question is therefore whether the strong link between the NSMCs and OSPAR continues to exist in the renewed sphere of authority.

Similar to earlier NSMCs, the issues discussed during the NSMC of 1995 were produced water, oil contaminated cuttings and chemicals (North Sea Ministerial Conference, 1995). The adoption of a harmonized mandatory control scheme was again asked for and OSPAR indeed adopted it in 1996. In 2002, there were again some interlinkages between the NSMC's political declaration and activities of OSPAR. The NSMC discussed some newer issues like hazardous components of produced water other than oil and the reduction of the volume of produced water discharges (North Sea Ministerial Conference, 2002). Although the OIC has discussed these issues (Secretariat of the OSPAR Commission, 2003), no concrete outcome has been developed yet.

The last NSMC, held in Gothenburg in 2006, however, only focused on fisheries and shipping and not at all on offshore installations. That means that since 2002, there has been no political impetus of the NSMC within OSPAR. Moreover, the 2006 NSMC was the last one in the sequence that started in 1984. The general feeling exists that the NSMCs no longer serve their purpose because other international organizations, i.e. mainly the IMO, OSPAR and the EU, have taken up the issues addressed by the NSMCs (North Sea Ministerial Conference, 2006). This means that the role of the NSMC in the renewed regional sphere of authority is different than in the initial sphere of authority. Important is that this change is not related to the policy changes discussed in this chapter, but to an overall increase in efforts and regulations for the environmental governance of the North Sea.

Second, environmental NGOs seem to have withdrawn their participation in the regional sphere of authority. To recall, the participation of environmental NGOs was related to the Brent Spar incident, which was a wake-up call to focus attention on the offshore industry. Second, the new rules of the game allowed observers to be present at meetings of the SEBA/OIC committee. However, Greenpeace has stopped attending these meetings in 2004 and during the 2008 and 2009 meetings no environmental NGOs were present at all. Although it is not entirely clear why these environmental NGOs have neglected recent meetings, the DTI expects (Interview Department of Trade and Industry, 2006c) that it is likely that these NGOs no longer find it necessary to put resources into combating environmental pollution from offshore oil and gas production and have moved on to other, more urgent, issues.

All in all, the environmental governance of the offshore industry is characterized by multiple spheres of authority. The OSPAR, national and EU sphere of authority co-exist, but also overlap and interact with each other. The industry self-governance, although not a separate sphere of authority is important within the environmental governance of offshore oil and gas production as well. Based on the previous Figures, the landscape of the environmental governance of offshore oil and gas production looks as follows (Figure 10).

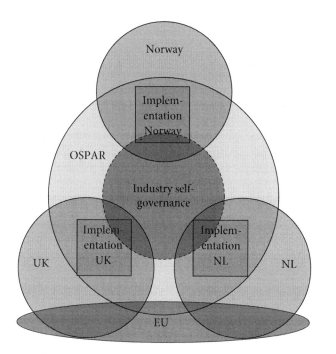

Figure 10. The spheres of authority in the environmental governance of oil and gas production.

5.10 Authority of the state during 2000s

The emergence of the EU sphere of authority and the subsequent renewal of the national spheres of authority means that who displays authority within the environmental governance of offshore oil and gas production has changed as well. Again, the changes in authority are related to the existence of multiple spheres of authority. The authority of the state is in the end a sum of its authority in these different spheres of authority. The changes in authority of the state discussed in this section are, however, not valid for Norway, because Norway is not affected by the emerging EU sphere of authority.

5.10.1 Authority in developing steering mechanisms

To recall, it was concluded that after the emergence of the national spheres of authority and industry self-governance, that authority over the development of steering mechanisms had become shared. Even after two policy changes, however, the authority of states within OSPAR over the development of steering mechanisms remains high although certainly somewhat shared with the industry. This authority might, however, be compromised by the fact that since a few years, the North Sea states lack the powerful impetus of NSMCs that used to reinforce their positions. With the end of the NSMCs, North Sea states have lost their extra forum for consensus building and agenda setting for issues and steering mechanisms with regard to issues that affect the North

Sea environment. Whether this is the case, remains to be seen in coming years and will require an in-depth analysis that goes beyond the scope of this thesis.

The other change in authority within OSPAR is the authority of environmental NGOs. They have ceased coming to the OIC meetings, which means that their direct involvement has died out in recent years. The authority of environmental NGOs on the development of OSPAR steering mechanisms has therefore at best been a temporary phenomenon.

Yet, more important for the changing authority of the state is the emergence of the EU sphere of authority. This has had two contradictory effects. On the one hand, the governmental actors active in the offshore industry and the industry itself feel a distance with the EU, making it more difficult to get a grip on the decision making process within the EU. The EU is a huge organization, which can only be influenced indirectly, while DTI is directly sitting at the table at OSPAR (Interview Department of Trade and Industry, 2006c). Similarly, in the Netherlands it is the Ministry of Environment who is closely involved in developing European environmental legislation, and not the Ministry of Economic Affairs who regulates the offshore industry. Industry associations have tried to influence the EU, but have for example difficulties in finding out which person is the right person to talk to (Interview Consultant offshore industry, 2006). This thus implies a loss of authority for both the state actors that regulate the offshore industry and the offshore industry itself, because a sphere of authority emerged in which they are not a structural part of the actor dimension.

On the other hand, the development of EU legislation has led to an increase in national formal regulations that the industry is faced with, indicating an increase in the authority of the state. In the UK, DTI has developed several regulations specifically with the purpose of applying EU legislation onto the offshore industry. Moreover, these regulations require the industry to have permits for all kinds of discharges and emissions. In the Netherlands, the Ministry of Environment and the Ministry of Nature are implementing EU legislation in formal laws. The Ministry of Economic Affairs is deliberating with the offshore industry through the Environmental Covenant how to implement the EU legislation that is applicable to the offshore industry. In both cases, the authority of the state has increased nationally. The remarkable fact is, however, that both DTI and the Ministry of Economic Affairs are not entirely pleased with this development, because the increased authority is the result of EU legislation whose development they are hardly involved in. Furthermore, the increased authority of the state is one that is shared with new governmental actors (the environmental Ministries). This has, especially in the Netherlands, led to cracks in the monopoly of authority of the Ministry of Economic Affairs with regard to the environmental issues of offshore oil and gas production.

The authority of the industry, which it had gained because of participatory steering mechanisms nationally, has subsequently declined. The newest standards come from the EU and the industry is required to implement them. Although these new standards and their implementation are deliberated within different forums in the UK and the Netherlands, the industry is on the receiving end.

In other words, overall, the authority of the state seems to have grown, while that of offshore the industry has declined. At the same time, however, the authority of those state actors involved in regulating the offshore industry and the authority of the offshore industry have declined, because of their lack of participation in EU decision making. This lack of participation has the consequence that both state actors and the industry have hardly any authority in the development of EU steering mechanisms that are relevant for the offshore industry.

5.10.2 Authority in generating compliance

Whether the authority of the state in ensuring compliance has changed because of European legislation is a more complex question. That is because for the UK and the Netherlands, the impact of the EU on ensuring compliance differs. The UK has seen the development of a plethora of formal regulations as a means for implementing EU legislation. This has also meant that the industry needs several permits. These permits are closely related to the compliance mechanism of inspections. Consequently, in the UK, the authority of the state in ensuring compliance has increased at the expense of the authority of the industry, because it relies much more on inspections than it used to.

In the Netherlands, EU legislation is formally implemented through national law, but in reality is implemented through the Environmental Covenant. This means that compliance by the offshore industry is arranged through reporting and much less so through inspections. Through the company environmental plans and the annual reports, the implementation of EU legislation is monitored by all participants of the Environmental Covenant. And as noted before, in reporting the responsibility lies with the state for ensuring the quality of information reported and ensuring that reporting is done in a proper manner, while the industry is the one that has to gather all information.

The authority of the state has the potential to increase when the infringement procedure is actively used. The fact that the European Commission can start the infringement procedure means that national governmental actors can put extra pressure on the industry to comply. Again, because EU steering mechanisms have a broader scope and the Environmental Ministries are more involved in implementing EU legislation, it is not necessarily the governmental regulators of the offshore industry that gain from the infringement procedure. This will depend on the directive the infringement procedure is running for and the allocation of responsibilities on the national level between different governmental actors.

At the same time, when the infringement procedure is started and when it affects the offshore industry it means that the European Commission has become an actor in the environmental governance of offshore oil and gas production that suddenly has authority in ensuring compliance.

5.11 Future outlook

After discussing past developments and changes in the environmental governance of offshore oil and gas production, it is worthwhile to take a look at future developments as well. Two are especially noteworthy here, because they are expected to have considerable impact on the way the environmental impact of offshore oil and gas production is governed, including the authority of the state and the industry. The first change will be the end of the Dutch Environmental Covenant. This Covenant was agreed for the period 1995 - 2010. That means that in the near future, the Covenant will expire. The question is whether this Covenant will be followed up by a new one or not. The general expectation is that the Environmental Covenant will not be renewed. There are several reasons for this. First, according to the industry (Interviews Gaz de France, 2007; Wintershall

Noordzee B.V., 2006), the industry is regulated and behaving well, a lot of targets are met and new targets would mean having to make investments that are not cost-effective. Second, the Ministry of Environment has not indicated that it wants to continue with and renew the Environmental Covenant. There is a general wish to continue the cooperation between the government and the industry, but the exact form is still unclear.

The second change is the development of the EU Integrated Maritime Policy. The EU launched their vision at the end of 2007. The EU is looking for a way to develop a policy with an integrated approach towards the different uses of the sea, including offshore oil and gas production, to ensure a sustainable development of the European seas. The EU is currently operationalizing its vision into concrete plans and policies. This means that in the near future the environmental governance of offshore oil and gas production, which was sector based until now, will broaden in scope as a result of EU legislation.

These two developments have the potential to considerably change the political landscape in the environmental governance of offshore oil and gas production. I even expect that the two developments will reinforce each other. With the end of the Environmental Covenant, a voluntary steering mechanism that diffuses responsibility and authority from the state to the industry will cease to exist. At the same time, the development of EU legislation – developed with little involvement of the actors currently involved in the spheres of authority within the environmental governance of offshore oil and gas production – will reverse the shift in authority to the industry back to other state actors. In other words, the offshore industry will lose some of its authority and influence onto the development of steering mechanisms it currently has to (other) state actors.

Besides the loss of authority for the industry, the Dutch national sphere of authority will also lose its importance, because it will lose its defining feature, the Environmental Covenant. Instead, the Dutch national sphere of authority, just like the UK sphere of authority already experienced, will move to a sphere of authority that is doing not much more than implementing regional steering mechanisms. At the same time, the EU sphere of authority will become increasingly important, because it will continue to develop legislation that is applicable to the offshore industry.

In other words, the complex situation that exists today in the environmental governance of offshore oil and gas production will become less complex. The regional spheres of authority (i.e. OSPAR and the EU) will be the ones developing steering mechanisms, while the UK and Dutch sphere of authority will shrink and overlap more and more with the regional spheres of authority. The only exception will be Norway, which is continuing with its zero discharge approach and which is to a lesser extent affected by the increasing importance of the EU sphere of authority.

Chapter 6.
Comparing the environmental governance of shipping and offshore oil and gas production

6.1 Introduction

The previous two chapters showed how the environmental governance of shipping and offshore oil and gas production developed and changed over the years. For both sectors, it was analysed how the traditional sphere of authority emerged, what the authority of the state was in this sphere of authority, how the environmental governance of both sectors changed through the emergences of new spheres of authority and the renewal of existing spheres of authority, and how that caused changes in the authority of the state. In this chapter, comparisons between these two case-studies will be made in those areas, although first a comparison will be made of the nature of the industrial activities of shipping and offshore oil and gas production. This comparison allows for a better understanding of the similarities and differences in the emerged spheres of authority in the environmental governance of shipping and offshore oil and gas production.

In making these comparisons, four of the research questions posed in Chapter 3 will be answered:

- Which policy changes have taken place since the emergence of the first sphere of authority in the environmental governance of shipping and of offshore oil and gas production?
- What is the effect of these policy changes in terms of the renewal of existing and the emergence of new spheres of authority in the environmental governance of shipping and of offshore oil and gas production?
- To what extent has the renewal of existing and the emergence of new spheres of authority led to shifts in governance in the environmental governance of shipping and of offshore oil and gas production?
- How has the authority of the state in developing steering mechanisms and generating compliance changed as a result of the policy changes in the environmental governance of shipping and of offshore oil and gas production?

6.2 The nature of maritime activities

The main similarity between commercial shipping and oil and gas production on platforms is that both take place in the marine environment. Both activities are connected with land, in shipping because of their visits to ports and in oil and gas production because of pipelines. Yet, an important part of the operations concerning shipping and oil and gas production take place at sea and thus sometimes also outside national borders. The environmental flows of discharges and emissions thus also take place at sea. Oil and chemicals enter the marine environment while air emissions enter the atmosphere. What is more, this pollution crosses national borders.

In shipping, this cross-boundary nature of pollution is amplified, because the ship itself crosses borders during its voyage. Shipping is a truly global activity. Shipping consists of flows of ships, goods and pollution around the world. Moreover, the activity of shipping is for a part done by companies who operate worldwide. This global nature of both the activity and its pollution explains why the environmental governance of shipping is strongly characterized by international decision making processes and steering mechanisms.

In contrast, platforms are static. Platforms are built in one place and stay there until they are decommissioned again. Sometimes floating platforms exist, which can be moved. In both cases, however, platforms are related to a field from which the oil and/or gas is extracted. Several platforms are constructed to cover one field. Moreover, both the fields and platforms are within national boundaries. Offshore oil and gas production takes place within the territorial sea or within the exclusive economic zones that coastal states can and have established. Governing offshore oil and gas production is therefore much more nationally organized than shipping.

Still, even though the platforms themselves are static and placed within national boundaries, the activity of oil and gas production also has transnational dimensions. For example, platforms are scattered around the North Sea because oil and gas fields exist in the entire North Sea area. Similar to the Gulf of Mexico or Western Africa, offshore oil and gas production is not limited to one country, but takes place in a group of countries that share oil and gas resources. Moreover, many companies involved in offshore oil and gas production are multinational and operate in several of such regions. These transnational characteristics come back in the regional convention that is at the heart of the environmental governance of offshore oil and gas production in the North Sea.

One of the main differences between the two activities is the history that they have. Shipping is a very old practice. Commercial shipping already takes place for many centuries. Over the centuries, ships have of course expanded in size, types, speed, etc. More importantly for this thesis is, however, that this long history comes back in the environmental governance of shipping in several ways. First, because of the freedom of the sea principle that has been governing the use of the sea by ships since the 17th century. Associated with that, flag states have a very important role in giving a nationality to and a home for a ship that is most of its time out at sea. Finally, the International Maritime Organization was already established in the 1940s; only three years after the United Nations was founded.

In contrast, offshore oil and gas production is a relatively young activity. Even though commercial oil extraction already exists since the 1850s, it has always taken place on land. Offshore oil and gas production started in the 1940s in the Gulf of Mexico, and in the North Sea at the end of the 1960s. The start of oil and gas production in the North Sea was followed-up with the 1974 Paris Convention to cover environmental aspects relatively quickly. This means that despite the fact that shipping is a much older sea-based activity than offshore oil and gas production, both sectors have a relatively long history in being subject to transnational environmental governance.

6.3 The evolution of the environmental governance of shipping and offshore oil and gas production

This section will, after having compared the traditional spheres of authority, answer two research questions of this thesis: which policy changes have taken place since the emergence of the first sphere of authority in the environmental governance of shipping and of offshore oil and gas production? And what is the effect of these policy changes in terms of the renewal of existing and the emergence of new spheres of authority in the environmental governance of shipping and of offshore oil and gas production?

6.3.1 Traditional spheres of authority

The similarities and differences in the nature of the activities of shipping and offshore oil and gas production help to understand the way in which the traditional spheres of authority in the environmental governance of shipping and of offshore oil and gas production are organized. For example, in both cases, the traditional spheres of authority evolved around a transnational forum that developed transnational steering mechanisms, i.e. respectively the International Maritime Organization and the Paris Convention. These transnational forums were very specialized forums with participating actors that were closely involved in both sectors. The emergence of transnational spheres of authority is undoubtedly influenced by the transnational characters of these two marine activities.

Yet, despite the fact that the nature of the activities might show some important differences, the characteristics of their traditional spheres of authority were remarkably similar. In both spheres of authority, the development of steering mechanisms took place within the transnational body. This included relatively formal rules of the game in the form of the IMO and Paris Convention that stated rules such as the procedures for the adoption of steering mechanisms and the tasks and responsibilities of different bodies and actors.

This also included states being the main actors in both forums because they have the formal authority to develop steering mechanisms; states were Parties to the Conventions, the voting members in IMO and Paris bodies and because states had jurisdiction over ships flying their flags and over the platforms in their territory. This, however, does not mean that other actors were not involved in the early days of the environmental governance of shipping and offshore oil and gas production. For example, industry associations have been part of the traditional spheres of authority. Within IMO, industry associations were present at meetings since at least the 1960s. In the Paris bodies, industry associations were not allowed to be present, but were allowed to present their views at the beginning of a meeting. In both sectors, lobbying the state actors was another way in which the industry was indirectly involved in setting standards for shipping and offshore oil and gas production.

Besides the activities of ACOPS to put the issue of oil pollution onto the agenda of some large flag states, both traditional spheres of authority were characterized by the lack of structural involvement of environmental NGOs. This might be explained by the fact that both spheres of authority emerged in an era in which environmental NGOs were not a common phenomenon yet and the fact that both activities take place at sea where pollution was not visible to and not directly experienced by the wider public.

In terms of power resources and relations, the states have always had power because they have the formal authority to adopt steering mechanisms. However, industrial actors have always had an important power resource because they have knowledge and expertise about industrial practices and in particular technologies that are an important instrument in reducing discharges and emissions.

Finally, implementation of and ensuring compliance with these steering mechanisms, however, took place on the national level. Within shipping, this distinction was more clear-cut, because the IMO did not have any compliance mechanisms. In contrast, the Paris bodies had several compliance mechanisms of which the main one is reporting. Still, in both traditional spheres of authority, it is the state that had the formal responsibility of implementing the transnational regional steering mechanisms. The authority of the state in implementation was, however, always compromised by the dependency on industrial actors to comply, especially with the lack of compliance mechanisms that existed in the traditional spheres of authority. This has, however, been more so in the case in shipping than in offshore oil and gas production where reporting was already a common compliance mechanism.

Besides such similarities that make the first spheres of authority in the environmental governance of shipping and of offshore oil and gas production to a large extent comparable, there was also one interesting difference between shipping and oil and gas production: the environmental governance of shipping is very much driven by incidents. Accidents like the Torrey Canyon (1967), Amoco Cadiz (1978), Erika (1999) and Prestige (2002) have led to acceleration in the adoption of new steering mechanisms. The environmental governance of offshore oil and gas production is hardly driven by incidents, but rather by results from monitoring practices. The only exception has been the Brent Spar incident.

6.3.2 Policy changes and the emergence of new spheres of authority

The first policy change in the environmental governance of shipping took place in the 1970s and 1980s. Driven by the adoption of the MARPOL Convention and the development of the UN Convention on the Law of the Sea, the rights of coastal and port states were defined and introduced. The Law of the Sea imposed new rules of the game to the IMO sphere of authority. The result was an increase in authority for coastal and port states. The changed institutional structure of IMO sustained that change.

In contrast to that, the first policy change in the environmental governance of offshore oil and gas production has been the emergence of new spheres and source of authority. National spheres of authority emerged, while at the same time the industry developed a system of self-governance through industry steering and compliance mechanisms. The emerged national spheres of authority in the UK, Norway and the Netherlands were influenced by the self-governance initiatives of the industry. All three national spheres of authority are characterized by the evolvement of joint initiatives between governmental actors and the industry. These joint initiatives were made possible, because the industry and the governmental actors recognized the need for improving environmental performance through involvement of the industry in decision making and implementation processes. The result of this policy change has been that the regional sphere of authority was no longer the only sphere of authority in the environmental governance of offshore oil and gas production.

Besides these very different policy changes, the environmental governance of shipping and offshore oil and gas production have also undergone a comparable policy change. In both sectors, the European Union emerged as a new sphere of authority. In shipping, this started in the early

1990s, although the influence of the EU on shipping was at its peak in the early 2000s. In offshore oil and gas production, this change is a phenomenon of especially the late 1990s and 2000s.

Even though there is similarity in the emergence of the European Union as a sphere of authority, there are also important differences in why and how this policy change occurred in the environmental governance of both sectors. As to the why, the focus of the EU on shipping was induced by the increase of substandard ships that traded with the EU and by the Erika and Prestige tanker accidents that polluted beaches of especially Spain and France. The EU subsequently developed a policy strategy for shipping. In contrast to that, for offshore oil and gas production, this policy change has been incremental. The EU does not have a clear policy strategy for regulating the offshore industry, but has been regulating the industry by including them in directives focused on a variety of issues and sectors (nature conservation, air pollution, chemicals, etc.). The reason why has to be sought in a more general trend of an increasing focus on the protection of the marine environment by the EU, rather than in a specific focus on offshore oil and gas production.

Another difference is how this policy change has played out in the environmental governance of shipping and offshore oil and gas production. In shipping, the EU has caused a new dynamic between the EU and IMO and within the IMO sphere of authority. However, this new dynamic only occurs on certain issues, i.e. double hulls in the past and air emissions more recently. This means that the EU is not a structural sphere of authority in the environmental governance of shipping, but comes up in certain issues. On top of that, the EU is also not a structural sphere of authority, because so far, the IMO has integrated the deviating EU steering mechanisms within its own sphere of authority.

In contrast, in the environmental governance of offshore oil and gas production, the EU has become a structural and complementary sphere of authority. Thus the environmental governance of offshore oil and gas production consists of the regional sphere of authority around OSPAR, the EU sphere of authority, national spheres of authority in the UK, Norway and the Netherlands and the industry self-governance. This means that even though the regional sphere of authority has remained fairly stable, the central position of this sphere of authority is in decline.

The first conclusion that can be drawn from this comparison is that the environmental governance of shipping and offshore oil and gas production have changed differently, despite great similarity in the traditional spheres of authority. That is, the first policy changes were different, i.e. UNCLOS versus national spheres of authority and industry self-governance as new sources of authority. The second policy change, the emergence of the EU, was similar but differed as to why and how the policy change occurred.

Second, despite these differences in policy changes, it can be concluded that changes in both the environmental governance of shipping and the environmental governance of offshore oil and gas production have been caused by external developments such as the adoption of UNCLOS and the application of EU directives on nature conservation offshore. This does not mean that there have not been any internal changes in the environmental governance of shipping and offshore oil and gas production (think for example about the environmental NGOs that have become an actor in both cases). However, it does mean that such internal changes have not resulted in a renewed sphere of authority. In sum, substantial changes in environmental governance are thus most likely to be related to developments that occur outside of the existing sphere(s) of authority.

6.3.3 Renewed spheres of authority

The first policy change in the environmental governance of shipping, i.e. the rights and obligations that were established for port and coastal states through UNCLOS, meant a breakdown of the central position of the flag state in the renewed global sphere of authority. The new rules of the game that increased the rights of port states explain why the IMO sphere of authority renewed itself. In a way, the authority of a port state to set standards for all ships entering its ports competes with the authority of flag states to set standards for its ships. A flag state can set (international) standards for its ships, but if one of its ships goes to a port where a different standard is required, then that ship has to meet that requirement or choose not to enter that port. The same goes for ensuring compliance in the renewed sphere of authority. A port state can inspect a ship and enforce compliance with port and international standards regardless of the origin and flag state of the ship.

Because of the distribution in the authority to set standards and enforce them between flag and port states, a new power mechanism became part of the renewed global sphere of authority as well. This power mechanism is unilateral action. Unilateral action by a flag state was against the nature of shipping and of being a flag state, because that creates competitive disadvantage which would lead to ships registering in other countries (i.e. flags of convenience). Port states, however, are more susceptible to unilateral action, because they are often the victims of pollution of foreign ships. Moreover, their competition is much more regionally based. For example, the US and the EU have set standards as a (group of) port state(s). With these unilateral or regional actions, the much desired level playing field and uniform standards of IMO is under pressure.

Other characteristics of the renewed sphere of authority have been an increased environmental focus of IMO and improved compliance mechanisms. The first is partly caused by the institutionalization of the port and coastal states that have higher environmental interests. As a result, the rules of the game that set down the mission of IMO changed and now also include the prevention of pollution from shipping. It can also be partly explained by the environmental NGOs that have become active in the sphere of authority and which have strengthened the call for environmental standards. Finally, it should be noted that the compliance dimension improved because of the establishment of an adequate system of port state control through the regional MoUs of Port State Control.

Even though this policy change caused the renewal of the IMO sphere of authority, certain elements in this sphere of authority remained stable. The main ones are the central position of the IMO in the environmental governance of shipping and the discourse that protecting a level playing field through global standards is the only appropriate way to deal with environmental issues in shipping.

The second policy change that affected the environmental governance of shipping has been the emergence of the EU as a source of authority. The IMO regards the EU as a competing source of authority in those cases where the EU sets standards that overlap with the scope of IMO and at the same time deviate from the IMO standards. Examples are that the EU decided to set different dates for the entry into force of double hulls and more recently the threat of the EU to develop legislation to limit air pollution when the IMO would not develop stronger limits.

This sense of competitiveness is felt very strongly in IMO, because global regulations are regarded as the only legitimate way of dealing with pollution from shipping. The EU steering mechanism threatened the level playing field that is advocated by most flag states and the industry. That also explains why the EU has had a considerable impact on the global sphere of authority. If the EU puts its weight behind an issue, the European actors force IMO to adopt the EU desired standards. This in turn caused changes in the power dimension, which then includes the EU coalition expressing a voice for the environment.

Thus, the EU policy change does affect the IMO sphere of authority, but only in specific issues. What should also be noted is that in these issues where the EU has been showing its teeth, the IMO has been able to restore the level playing field by accommodating the interests of the EU trough the development of global steering mechanisms with standards that satisfy the EU. In doing so, the IMO has so far always been able to end the competition with the EU and resume 'business as usual'. In addition, in the areas where the EU does not develop its own standards and does not enter into a competitive struggle with the IMO, the IMO sphere of authority as it renewed after UNCLOS marks the environmental governance of shipping. It can therefore be concluded that the IMO sphere of authority is still the most important source of steering mechanisms.

The only area in which the EU has caused structural change is in the area of implementation and compliance. In those cases, where the EU has adopted standards for shipping, the implementation of these standards is more strictly controlled. The member states are subject to the infringement procedure when they do not implement EU legislation properly. Consequently, because EU and IMO standards overlap on certain issues, the implementation of IMO standards by EU member states has improved as well. In addition, the EU has made port state control, an important compliance mechanism in the global sphere of authority, stricter.

In sum, the two policy changes that have occurred in the environmental governance of shipping have changed the IMO sphere of authority, but have not led to new spheres of authority. Only the EU can in some issues and for some periods (i.e. until IMO copies the EU standards) be regarded as a separate sphere of authority. Changes in the environmental governance of shipping have thus been caused by external developments that have led to the renewal of the IMO sphere of authority rather than to the emergence of new spheres of authority.

In contrast, the effect of the two policy changes on the environmental governance of offshore oil and gas production is very different. After two policy changes, the environmental governance of offshore oil and gas production no longer consisted of one sphere of authority, but of multiple. First, there are the national spheres of authority consisting of joint initiatives between governmental actors and the industry and a discourse on stakeholder dialogue. Second, there is industry self-governance with industry steering and compliance mechanisms. Third, there is the EU sphere of authority with EU steering and compliance mechanisms. Finally, the traditional regional sphere of authority continued to exist as well, albeit in a renewed form.

One of the main effects of the first policy change (i.e. the emergence of national spheres of authority and industry self-governance) on the regional sphere of authority is that rules of the game now allow observers to be present at meetings of the working group that deals with offshore installations (SEBA and later OIC). In addition, environmental NGOs have become part of the regional sphere of authority. Yet, while the environmental NGOs participation has died out in recent years, the industry associations have become a valued addition to the meetings, because of their cooperative attitude and knowledge and expertise. This also explains why the discourse and practice of stakeholder dialogue, which already characterized the national spheres of authority, has penetrated the regional sphere of authority as well.

Another aspect of the renewal of the regional sphere of authority has taken place in the results dimension. The influence of national spheres of authority and industry self-governance has changed the national context in which regional steering mechanisms are implemented. For example, the industry has gained a much bigger role. This change is caused by the improved deliberation structures that exist between the government and the industry within the national spheres of authority. That means that the industry is involved in the implementation of regional steering mechanisms from the outset. In the UK, the implementation on the national level is done through formal regulations, while the industry has to incorporate these new regulations into their own environmental policies and management systems.

In Norway and the Netherlands, the implementation of OSPAR decisions and recommendations is not only done through formal regulations but also through the joint initiatives of respectively the zero discharge policy and the Environmental Covenant. The industry has been made more responsible for the improvement of environmental performance under these joint initiatives. Moreover, some industry steering mechanisms have close ties with these national joint initiatives, because they are used to operationalize and implement regional and national requirements. This explains why the industry and its self-governance initiatives have become linked to the regional sphere of authority.

The policy change around the EU has not had much impact on the regional sphere of authority, i.e. the regional sphere of authority did not renew itself. That is because the EU regulates different issues than OSPAR does. There is therefore hardly any competition between the EU and OSPAR. The only area where they overlap is the issue of chemicals. The EU developed regulations on chemicals which also apply to the offshore industry. However, OSPAR had already developed similar steering mechanisms for offshore chemicals. So far, OSPAR has compared its own steering mechanisms with the EU one and has tried to establish contact with the EU on how to deal with the issue of chemicals and the overlap in OSPAR and EU regulations further.

However, the emergence of the EU has led to a renewal of the national spheres of authority. The UK and the Netherlands have been implementing EU steering mechanisms mostly by national law and regulations. Especially the UK has seen a growth of formal regulations for its offshore industry as a result of EU legislation. In the Netherlands, the implementation of EU legislation is not only arranged through formal regulations, but also through the Environmental Covenant. Moreover, in both countries, a distance is felt with the EU, because both the governmental actors regulating the offshore industry and the industry itself are only to a limited extent involved in the development of these EU steering mechanisms. Authority is thus moving away from the offshore regulatory actors and the offshore industry. The Norwegian sphere of authority has not been affected, because it is not a member of the European Union.

The more general conclusion that can be drawn from this comparison is that the differences in the environmental governance of shipping and of offshore oil and gas production have increased. The differences between both cases is explained by the different policy changes that affected the environmental governance of shipping and offshore oil and gas production. One of the biggest differences is that while the environmental governance of shipping continues to be subject to one central sphere of authority (and one extra sphere of authority that is relevant in a limited number of issues and for shorter periods of time), the environmental governance of offshore oil and gas productions knows several spheres and sources of authority. Thus, the IMO has remained dominant in shipping, while the regional OSPAR Commission and bodies have lost authority to the European Union, national governments and the industry.

The second conclusion is that the effect of policy changes is not that the initial or traditional spheres of authority disappear, become redundant or become insignificant. Rather, the traditional spheres of authority, i.e. the IMO and OSPAR, continue to exist and play an important role in the environmental governance of both marine activities. Thus, environmental governance is increasingly characterized by both traditional and new spheres and sources of authority.

Finally, this comparison also shows that these traditional spheres of authority do change as a result of policy changes. The traditional spheres of authority remain important in environmental governance but have renewed as a result of policy changes and other, internal developments. Thus even in the environmental governance of shipping, the organization, substance and results of the central sphere of authority are currently different from how they used to be in the 1950s, 1960s and 1970s. At the same time, however, some elements have remained traditional and continue to be part of the organization, substance and results dimensions. It can thus be concluded that not only does the environmental governance of both sectors increasingly consist of 'traditional' and new spheres of authority, but that 'traditional' spheres of authority in their renewed forms also consist of 'traditional' and new elements.

6.4 Shifts in governance

In Chapter 2, the changes in governance that are observed in the last two decades were related to four shifts in governance, i.e. the shift to multiple actors, levels, rules and steering mechanisms. In this section, I will review to what extent these shifts have also occurred within the environmental governance of the two marine activities central in this thesis. In other words, I will answer the following research question: to what extent has the renewal of existing and the emergence of new spheres of authority led to shifts in governance in the environmental governance of shipping and of offshore oil and gas production?

The first feature of changing governance practices has to do with a shift from spheres of authority based on a single actor to spheres of authority that include multiple actors. One of the first things that can be concluded in this regard is that the traditional spheres of authority were in fact not as single-actor based as governance literature suggests. Yes, the state played the most prominent role in the traditional spheres of authority in the environmental governance of shipping and of offshore oil and gas production, but especially industrial actors were part of them as well. At the same time, it is clear that the renewed and new spheres of authority also have multiple actors involved. The difference is that in these spheres of authority states still have a major role to play, but environmental NGOs and industrial actors have gained a more structural place. In other words, environmental NGOs and especially industrial actors have gained authority in environmental governance practices. The shift to multiple actors is thus not only about whether there are several types of actors involved, but also about how much authority they have. Looking at it from this perspective, I conclude that there has indeed been a shift from a few actors to multiple actors, because compared to the traditional governance practices more actors (especially in offshore oil and gas production) have gained authority in environmental governance of marine activities.

This shift to multiple actors can go hand in hand with the second shift in governance, the shift to multiple levels. However, in marine governance, or at least in shipping and offshore oil and gas production, multiple levels have always been a core feature. The environmental governance of both marine activities have from the start been characterized by transnational cooperation and national implementation. Over the years, the national level was strengthened in the environmental governance of offshore oil and gas production and in both cases the European Union forms an extra 'governance' layer. Thus, it can be concluded that there has indeed been a shift from two to more levels. However, it should be noted that the traditional multi-level characteristic of the environmental governance of shipping and offshore oil and gas production is a rare phenomenon.

The third shift in governance is the shift from a set of formal rules of the game to multiple and sometimes informal rules of the game. The assumption behind this shift is that new informal rules emerge when new governance practices come into existence. Traditionally, the environmental governance of shipping has been mostly embedded in formal institutions. Yet, this continues to be the case even in the renewed sphere of authority. The importance of IMO as the forum for developing steering mechanisms and the formal rules of the game that IMO provide have remained important for the environmental governance of shipping. Only with the emergence of the new EU dynamics, informal rules of the game (i.e. about how to deal with EU within the IMO) have emerged. The environmental governance of shipping is therefore only in a limited way affected by a shift to multiple rules.

In contrast, the environmental governance of offshore oil and gas production has undergone this shift to multiple rules. That is because new spheres of authority with new rules of the game have emerged (i.e. the Environmental Covenant, environmental management systems, etc.). These rules of the game have not always been informal, but were new and experience with and embedding of such rules takes time. Conclusions drawn from these observations are therefore that such a shift to multiple rules is indeed related to the emergence of new governance practices – often in new spheres of authority – and that new rules of the game are not necessarily informal but can be formal in nature as well. A difference in this regard should be made between rules of the game that are formal because they are embedded in formal documents and between rules of the game that have become formal because they already exist for a long time. It is the first type that is part of new spheres of authority, while the latter is much more restricted to the longer existing spheres of authority.

The final shift in governance that was mentioned in Chapter 2 is the shift from single steering mechanisms to multiple steering mechanisms. It is interesting to note that the three previous shifts are related to the organization and substance dimensions of spheres of authority in governance, while this one is related to the results dimension. Are the results of marine governance indeed increasingly based on several steering mechanisms instead of a single one? In shipping, the steering mechanisms within IMO, i.e. international conventions, continue to be the main steering mechanisms. Yet, if one takes a better look at these IMO steering mechanisms one trend has been a shift from technology-following to technology-forcing steering mechanisms. In addition, in the last decade, European ones have been added to these two types of IMO steering mechanisms. However, these IMO and European steering mechanisms overlap to a large extent. This consideration leads to the conclusion that the environmental governance of shipping has experienced only a very limited shift towards multiple steering mechanisms.

In contrast, this shift to multiple steering mechanisms has been much more important in the environmental governance of offshore oil and gas production. Besides the OSPAR steering mechanisms, European, national and industry steering mechanisms have emerged. It was also argued that most of these steering mechanisms complement each other. Based on these observations, it can be concluded that the shift to multiple steering mechanisms can be driven by two developments. This shift occurs either because several types of steering mechanisms evolve within one sphere of authority or because new spheres of authority with other actors and at other levels develop their own steering mechanisms.

Based on this analysis, the conclusion can be drawn that in cases where new spheres of authority emerge, it is more likely that all shifts occur. Similarly, if existing spheres of authority remain stable or renew as a result of a policy change, the shifts to multiple actors, levels, rules and steering mechanisms do not necessarily occur. They can occur, when new practices emerge within the existing sphere of authority. Although this conclusion seems to be very logical, it would be interesting to seek examples where changes in existing spheres of authority have led to these four shifts in governance, or whether new spheres of authority have only led to these shifts in very limited ways. This could help in broadening the understanding of when, why and how these shifts in governance emerge and how they affect governance practices.

6.5 Changing authority of the state

The next area of comparison is the changing authority of the state in the environmental governance of shipping and of offshore oil and gas production. Have the changes in governance discussed above had similar or different implications for the authority of the state? One would expect that the authority of the state in the environmental governance of shipping would change less than the authority of the state in the environmental governance of offshore oil and gas production, because the environmental governance of shipping has to a lesser extent been affected by the four shifts in governance and a subsequent diversification in spheres of authority. This section therefore also serves as the answer to the fourth research question of this thesis: how has the authority of the state in developing, implementing and enforcing policy changed as a result of the policy changes in the environmental governance of shipping and of offshore oil and gas production?

6.5.1 Traditional authority of the state

It was already discussed that state and industry actors formed part of the traditional spheres of authority in both shipping and offshore oil and gas production. Even though in shipping, an environmental NGO played a role in the development of the OILPOL Convention because it put the issue of oil pollution by shipping on the agenda, the influence of environmental NGOs on negotiations and the actual content of the steering mechanisms during the 1960s and 1970s was very limited. It was also argued that states were formally the most important actors in the traditional spheres of authority. The question in this section is thus whether this formal position of states is aligned with the authority of these states in the development and implementation of steering mechanisms and to what extent that differs for shipping and for offshore oil and gas production.

Whether this is the case, depends for a large part on the power struggle between the states and the industry over authority in the development and implementation of steering mechanisms. In the development of steering mechanisms, the states were the ones proposing new or changes to existing steering mechanisms. They were also the ones making the final decision to adopt a steering mechanism or not. This was the case in both the environmental governance of shipping and in offshore oil and gas production.

Moreover, it was a specific group of states that mattered most or had the most interest in the development of steering mechanisms. For shipping these were the flag states. In offshore oil and gas production, it were those states with platforms in their territorial waters or exclusive economic zones. Within the Paris Convention that were the UK, Norway, the Netherlands, Denmark, Germany and Ireland. The UK, Norway and the Netherlands were in turn the biggest producers of oil and gas.

However, in the mean time, the industry actors were trying to influence the development of steering mechanisms as well. In shipping, the industry introduced best practices and techniques as a means of influencing the negotiations. They developed these practices and techniques to circumvent expensive regulations and/or to reduce their own operation costs. Another way to exercise authority in developing steering mechanisms was by lobbying governments and by aligning with flag states. Because the interests of the shipping industry are similar to those of large flag states, it is difficult to assess the specific influence over governmental positions and the adopted steering mechanisms without detailed analysis. Unfortunately, this detailed analysis could not be done in this thesis. Still, what can be concluded is that the development of best practices or techniques was, for the industry, a powerful way to exercise authority during the negotiations by pushing for an alternative for stringent and expensive regulations.

In offshore oil and gas production, the industry associations also lobbied their governments and the group of governmental actors that met within the Paris Commission and Working Group on Oil Pollution, among others by giving presentations at the beginning of meetings. Their input and expertise was also explicitly sought by the Paris actors in certain cases, such as with oil based muds and cuttings. The role of the oil and gas industry is thus similar as that of the shipping industry when it comes to sharing expertise, knowledge and views about environmental effects and possibilities to reduce discharges.

All in all, the authority over the development of steering mechanisms was in both spheres of authority shared between the states and the industry. In the development of steering mechanisms there was a 'division of labour' between the states who proposed, negotiated and decided, while the industry provided expertise, practices and technologies to influence the proposals, negotiations and the final decision. Even though authority was shared, the authority was shared between the state and the industry unequally. The states had in the end decisive authority in the development of steering mechanisms.

This division of labour was different in the implementation of the transnational steering mechanisms. Formally, the authority of the state in implementing the transnational steering mechanisms and ensuring compliance was very clear. In both shipping and offshore oil and gas production, the state was responsible for implementation and arranging ways to ensure compliance.

In practice, however, this authority of the state was more complex, because even though states were responsible, it is the industry that in the end had to live up to the standards. Compliance with transnational steering mechanisms was therefore arranged step by step. States complied with transnational conventions when they had implemented transnational standards in national steering mechanisms. The industry, in turn, complied when it was actually living up to these standards.

One of the transnational compliance mechanisms that were part of both spheres of authority was reporting. States have to report implementation efforts and (non-)compliance to the secretariat of the Convention. However, in shipping, this compliance mechanism was rarely lived up to. This is because states did not sent in reports about their implementation and enforcement efforts. But also because a compliance mechanism that required ships to report their compliance was lacking. Moreover, assessing non-compliance was very difficult because ships are often out at sea and not in the ports of the flag state in which that ship is registered. This made it very difficult for flag states to assess the level of compliance of their ships.

Contrary to that, the reporting requirements of Paris bodies emphasized the reporting of actual discharges and thus of whether the standards were met and not of the implementation efforts taken by the states. In contrast with shipping, the reporting requirement was well lived-up to within the Paris Convention. States have been reporting the failure of platforms to live up to the 40 ppm target and the discharges of oil based muds and cuttings. Only with chemicals, the reporting took some years to get fully started up. The static nature of platforms made it easier to gather information about the discharges from platforms.

In shipping, additional compliance mechanisms were used to enforce international standards. This compliance mechanism consisted of inspections by the port state. Port states were allowed to inspect ships and their oil record book. There are, however, two reasons why this compliance mechanism did not work well. First, because it was very difficult to assess violations with discharge limits at sea through inspections in ports. Second, because even if the port state detected a violation, the flag state decided whether or not to prosecute the ship.

In offshore oil and gas production, the state also used additional compliance mechanism, namely inspections and the requirement of monitoring. Both are, however, not fully suited to find non-compliance. Inspections of discharges were difficult, because inspections always have to be announced beforehand. Moreover, monitoring does not necessarily detect non-compliance, but allows for an overall assessment of the effectiveness of measures.

In other words, even though the states were formally responsible for compliance, they depended on the industry to comply and to assess compliance. In shipping, the dependence on the industry is bigger, because it is a footloose sector. Moreover, the limits set were discharge limits which made assessing compliance through inspections difficult as well. It is therefore the nature of the activity and the nature of the steering mechanisms that made ensuring compliance in shipping such a difficult undertaking.

In offshore oil and gas production, states were also dependent on the industry, but in a different way. This dependency existed because the activity takes place at sea, which makes inspections difficult, but also because the state depended on the industry for the information of discharges. The big difference with shipping was that in oil and gas production these information flows did exist. In other words, it was possible to assess compliance of the industry at the national level and the Paris level.

It can thus be concluded that compliance was better arranged in the traditional sphere of authority of offshore oil and gas production than in shipping. It can also be concluded that the authority of the state in compliance in the environmental governance of offshore oil and gas production was much bigger than in the environmental governance of shipping. In addition, even though ensuring compliance was formally exclusively in hands of the state, in practice authority over ensuring compliance was shared with the industry. The state depended to a considerable extent on the industry, which should be willing to share information about their environmental performance. It is, however, important to understand that the nature of the activities also play a role in defining the compliance dimensions of both spheres of authority and the authority of the state in ensuring compliance.

In sum, because the traditional spheres of authority in the environmental governance of shipping and of offshore oil and gas production showed great similarities, it was expected that the authority of the state would be similar in both spheres of authority as well. For the development of steering mechanisms this was indeed the case. In both cases, the state and the industry had authority over the development of steering mechanisms, although the authority of the industry depended on whether they could offer viable alternatives for stringent standards. Yet, with regard to ensuring compliance, there was a difference between the authority of the state in shipping and in offshore oil and gas production. In shipping, the state could not live up to its formal responsibility, with the result that the industry had much authority over whether steering mechanisms were complied with or not. In offshore oil and gas production, the state had more authority over compliance issues, because the compliance mechanism of reporting was successful. This is explained by the differences in compliance mechanisms and the difference in the nature of the two marine activities.

6.5.2 Changing authority of the state in the development of steering mechanisms

This section compares how policy changes have affected the authority of the state in the development of steering mechanisms in the environmental governance of shipping and of offshore oil and gas production. Even though the initial authority of the state was similar in both sectors, the differences in policy changes that have affected both sectors result in the expectation that the authority of the state has undergone a different development in the environmental governance of shipping than in the environmental governance of offshore oil and gas production.

One of the main changes affecting the authority of the state in the renewed global sphere of authority of shipping is the fact that formal authority has become shared between flag and port states. Despite this formal authority, the authority of the flag and port state actually depends on the power struggle between each other, but also with coastal states, environmental NGOs and industry actors. All these actors are present during IMO meetings and together debate the adoption or change of steering mechanisms. The precise authority of these actors varies per issue and in time and can only be assessed through in-depth research into the negotiations and the adoption of a specific steering mechanism. Still, several observations can be made about which factors are important in how authority is shared between these actors.

First, port states are able to exercise authority during negotiations because of the interplay between new rules of the game (i.e. UNCLOS), a new power mechanism (i.e. the threat of unilateral action) and the continued discourse of a level playing field. The formal authority that is laid down in UNCLOS means that the threat of unilateral action is 'real'. In addition, the discourse of level playing field means that the actors within IMO will take almost any effort to ensure consensus and the adoption of IMO steering mechanisms that undo the need for unilateral action. This explains why ports states have considerable authority in the development of steering mechanisms in the global sphere of authority.

Second, this interplay is also at the core of the authority of the EU within the global sphere of authority. In cases where the EU functions as a source of authority by taking the lead in the development of regional steering mechanisms, the EU challenges the dominant discourse by threatening with the development of EU steering mechanisms. Yet, it should be noted that the EU does this as a coalition and not necessarily as individual states. The influence of individual member states is dependent on the decision making process and the negotiations within the EU. There, the standards are laid down in European legislation, which is then the basis of the position that the EU as a coalition will advocate within IMO. This coalition on the one hand empowers the EU member states in achieving a desired output from IMO, but on the other hand limits the individual EU member states in the position that they can take within the IMO debate.

Finally, the shipping and oil industry have always been able to exercise authority through the development of technologies and best practices. However, a trend of the last decade is that standards are set that go beyond existing technology. This trend might be explained by the increased power of the environmental voice within the global sphere of authority. The environmental coalition consists of port states, coastal states and environmental NGOs. Since the authority of the port state is considerable, the push for environmental stringent steering mechanisms has grown. On top of that, with the issues of ballast water and TBT in anti-fouling paints there was already an increased sense of urgency, which has strengthened the call for ambitious standards. This increased authority of the environmental coalition has also meant that the industry loses some of its authority in the development of steering mechanisms. The industry keeps insisting on feasible standards that are achieved with proved technologies. However, since they do not always have proven technology to allow for feasible but ambitious standards, they are overthrown by the environmental coalition.

It can thus be concluded that there are several changes in the authority of the state within the environmental governance of shipping with regard to the development of steering mechanisms. First, authority is increasingly shared between different types of states. It is no longer the flag states that have (formal) authority, but port and coastal as well as EU member states have gained authority. Second, the authority of these states is no longer only shared with industry actors but also with environmental NGOs, who have become a structural actor within IMO. Third, the industry seems to have lost authority because of the decreasing relevance of proven technology in setting the standards.

In the environmental governance of offshore oil and gas production, the authority of the state no longer only depends on its authority within the regional sphere of authority alone, but also on its authority within the development of national, EU and industry steering mechanisms. Within the renewed regional sphere of authority, the authority of the state depends on the power struggle with the industry actors. Contrary to shipping, however, the authority of the industry has improved. The increased authority of the industry is explained by the interplay between the power mechanisms of knowledge and expertise and the discourse of stakeholder participation. Furthermore, this power mechanism has become stronger, because the industry develops industry steering mechanisms and changed its attitude towards environmental regulations. The discourse of stakeholder participation is in turn caused by the emergence of national spheres of authority that rely on stakeholder participation as well.

What has also changed in the authority of the state in the environmental governance of offshore oil and gas production is that states have an extra forum for the development of steering mechanisms, namely nationally. Yet, in those national spheres of authority, the authority to develop national steering mechanisms is shared between the governmental and the industry actors. Again, the changed attitude of the industry and the industry self-governance are at the base of the joint national approaches, which give the industry clear responsibilities in developing the steering mechanisms. In other words, the authority of the industry is based on rules of the game that lie down industry's responsibilities and that arrange structural deliberation between the government and the industry in developing steering mechanisms.

The increased authority of the industry in the environmental governance of offshore oil and gas production is rooted in the self-governance that the industry developed. This self-governance does not only mean a separate source of authority in which the industry has most authority (i.e. the state also has some authority in industry self-governance because regional and national requirements are the basis of industry environmental policies), but is also the basis for the increased authority in the regional and national spheres of authority itself. The interplay between the industry self-governance, the regional and national spheres of authority means a shift from authority from the state to the industry.

At the same time, the EU sphere of authority has led to a loss of authority of both state actors that are directly involved in regulating offshore oil and gas production and the offshore industry it self. This loss of authority is explained by the lack of participation of these state actors and the offshore industry in the EU sphere of authority. Since the directives were either already developed before it was decided to apply them offshore, or are broader than the offshore industry alone, the influence of the regulatory actors and the industry has been very limited. In fact, the development of steering mechanisms in the EU has resulted in a shift of authority to other state and private actors, because they are more involved in the decision making in the EU then the offshore states and industry.

In sum, the authority of the state in the environmental governance of offshore oil and gas production has changed in several ways. First, the industry has gained authority within the regional sphere of authority, the national sphere of authority and through its self-governance. Second, the authority of the state has gained a new dimension nationally. Yet at the same time, the states and the industry have both seen a loss in authority when it comes to the European sphere of authority. Overall, it can thus be concluded that the authority of the state in the development of steering mechanism has declined, while the authority of the industry in the environmental governance of offshore oil and gas production has increased.

From this comparison, we can conclude that the authority of the state in developing steering mechanisms has changed differently in the environmental governance of shipping than in offshore oil and gas production. This is explained by the differences in policy changes and the subsequent dynamics in the sphere(s) of authority in the environmental governance of both sectors. In shipping, the state remains the actor with most authority, although this authority is shared between different groups of states. In contrast, in the environmental governance of offshore oil and gas production, there has been a shift in authority away from the states to the EU and to industry actors. In other words, the authority of the state in developing steering mechanisms is larger in the environmental governance of shipping than in the environmental governance of offshore oil and gas production, where the authority of the state is shared with another sphere of authority (i.e. the EU) and with the industry.

6.5.3 Changing authority of the state in generating compliance

This section reviews the differences and similarities in the changing authority of the state in ensuring compliance in the environmental governance of shipping and offshore oil and gas production. The initial authority of the state in generating compliance within the environmental governance of shipping was less than that of the state in the environmental governance of offshore oil and gas production. The question is whether these differences still prevail after the policy changes both sectors have gone through?

The first observation that should be made is that the compliance dimension in the environmental governance of shipping has seen many changes and improvements. In contrast, in the environmental governance of offshore oil and gas production, the main change is not the way in which compliance is ensured (i.e. through reporting and inspections), but rather the involvement of the new spheres and sources of authority in ensuring compliance. The question remains how has this affected the authority of the state in both the environmental governance of shipping and of offshore oil and gas production?

Many of the improvements in compliance in the environmental governance of shipping are a direct result of the extended rights of port states to inspect, detain and prosecute ships visiting their ports. These port state inspections have been harmonized and have become a common practice through the regional MoUs on Port State Control. Moreover, the EU has strengthened the MoU on Port State Control in Europe even further. The effectiveness of these port state inspections has been further enhanced by the adoption of equipment standards as an addition to discharge standards. Equipment standards are much easier to inspect than discharge standards, because equipment is visible on a ship and discharges made or not made are not. This means that the authority of port states in generating compliance has increased extensively. In other words, the interplay between new rules of the game and a change in the type of steering mechanisms set has resulted in an effective compliance mechanism in the global sphere of authority and thus in more authority for the port state.

However, other actors have become involved in generating compliance as well. First, the EU has the infringement procedure through which the European Commission can ensure compliance with various IMO steering mechanisms. Second, the flag state has gotten more authority in ensuring compliance as well. IMO steering mechanisms require ships to carry several certificates. These certificates are only issued when a ship complies with international standards. Even though the flag state is formally issuing these certificates, in practice this task is conducted by classification societies. Thus classification societies, on behalf of flag states, play a role in ensuring compliance as well. Finally, the IMO has developed a voluntary compliance mechanism, i.e. the Voluntary Audit Scheme through which flag states can get their implementation efforts of IMO conventions voluntary audited.

In sum, in the environmental governance of shipping, efforts to improve compliance have grown tremendously resulting in authority in compliance for a variety of actors. These efforts have especially empowered state actors in controlling compliance. More specifically, this increased authority is shared between flag and port states, although port states with their system of port state control seem to be most equipped to detect non-compliance. But authority over compliance is still not limited to states. Other actors, i.e. the European Commission, classification societies and to a very limited extent IMO, have gained authority as well. As a result, the authority that the industry used to have because of the lack of governmental compliance mechanisms has significantly reduced.

In contrast, the environmental governance of offshore oil and gas production knows only one new compliance mechanism, the infringement procedure of the EU. However, this compliance mechanism is much less used as the two 'traditional' compliance mechanisms of reporting and inspections. What has changed instead is the extent to which they are used and, more importantly, who is employing these compliance mechanisms.

First, reporting has become more important and is no longer limited to the regional sphere of authority only. Rather reporting is spread throughout the regional and national spheres of authority and is an industry compliance mechanism as well. Reporting efforts are used economically, because reports of the industry on their environmental performance are used for their environmental management systems, their annual environmental report, for the national policies and joint approaches and for OSPAR. The more widespread use of reporting does not change the authority of the state. The states still set requirements for the quality of information gathered, and use this information to review compliance and the effectiveness of steering mechanisms. In addition, the dependency on the industry for the provision of information also continues to exist. However, the fact that the industry is also actively using reporting internally and as a way to communicate to the wider public results in an increased authority of the industry in ensuring compliance. Subsequently, the authority of the state in generating compliance has not necessarily changed, but is subject to a relative decline.

The use of the compliance mechanism of inspections has increased as well and with that the authority of the state. The increased use of inspections is explained by the emergence of the EU sphere of authority which has led to an increase in formal national steering mechanisms that require platforms to carry permits. Yet, the industry does help the state in doing these inspections, because the internal environmental policies and environmental management systems ensure an integrated approach to implementation of national and regional steering mechanisms which makes it easier for inspectors to control compliance with environmental standards when they inspect platforms. Second, the industry is conducting internal audits and is sometimes externally audited by certification bodies. The latter shows that the industry has gained some authority in ensuring compliance through inspections as well.

In sum, in both the environmental governance of shipping and of offshore oil and gas production the efforts and authority in generating compliance have increased. One of the differences is that in shipping, the number of actors having this authority has grown, while in offshore oil and gas production, generating compliance remains limited to the state and the industry. Moreover, in shipping the authority of the state in generating compliance has increased considerably, while in offshore oil and gas production this authority is much more shared with the industry.

Chapter 7.
Conclusions and reflection

7.1 Introduction

This thesis started with the introduction of the marine environment as an interesting policy domain for a study into governance practices. The marine environment and its special characteristics would serve as a breeding ground for governance. The sea knows many users with each their own stake in using the sea, decreasing sovereignty for states the further one gets away from the coastline and maritime activities generating transboundary pollution problems. It was expected that these characteristics would induce the development of innovative (transnational) governance arrangements.

The issue of governance has been subject to debate by social scientists for at least two decades. However, social scientists still seem puzzled about what governance is, how one should analyse governance practices and what the implications are of shifts in governance for the authority of the nation state. Taking the marine environment as a policy domain, this thesis explored exactly these issues. In particular, because of their long history in environmental governance, shipping and offshore oil and gas production were the two cases in which shifts in governance and the subsequent changes in the authority of the state were analysed in this thesis. Moreover, the North Sea was chosen as the region in which the environmental governance of these maritime activities was analysed, because of the plethora of activities, stakeholders and spheres of authority that exist in this region. The objective of this thesis therefore was *to analyse how shifts in governance affect and change the authority of the state in the environmental governance of shipping and of offshore oil and gas production in order to generate new insights for the governance debate.*

In order to analyse the environmental governance of shipping and offshore oil and gas production and the changing authority of the state, a conceptual framework was developed in Chapter 2. The concept sphere of authority is at the heart of this conceptual framework and was defined as *the temporary stabilization of the organization and substance of a policy domain within which actors take decisions through the development of steering mechanisms and/or have the ability to generate compliance with these steering mechanisms.* The dimensions actors, power relations, rules of the game, discourse, steering mechanisms and compliance mechanism allow for an analysis of the organization, substance and results of a sphere of authority. Based on this, the authority of the state, or other actors, in developing steering mechanisms and in generating compliance within the sphere of authority can be assessed. The term policy change allows for analysing how new spheres of authority can emerge and/or how existing ones can renew itself.

Based on this conceptual framework, the research questions of this thesis were defined as follows:

- Which policy changes have taken place since the emergence of the first sphere of authority in the environmental governance of shipping and of offshore oil and gas production?
- What is the effect of these policy changes in terms of the renewal of existing and the emergence of new spheres of authority in the environmental governance of shipping and of offshore oil and gas production?

- To what extent has the renewal of existing and the emergence of new spheres of authority led to shifts in governance in the environmental governance of shipping and of offshore oil and gas production?
- How has the authority of the state in developing steering mechanisms and generating compliance changed as a result of the policy changes in the environmental governance of shipping and of offshore oil and gas production?
- What are insights gained from the development and application of the theoretical concepts of this thesis and the results of the empirical analysis for the governance debate?

Chapter 4, 5 and 6 answered the first four research questions. This chapter will answer the fifth research question, which asks for the new insights for the governance debate generated by this thesis. However, before exploring the new insights generated by this thesis, this chapter will first reflect on what can be learned from this thesis with regard to the policy domain of the marine environment. Is governance at sea indeed a breeding ground for new and innovative ways of governing? After the theoretical reflections, this chapter and thesis will end with an epilogue. In this epilogue, I will express my ideas about the future of the environmental governance of maritime activities. Can the environmental governance of shipping continue with depending on a single sphere of authority? And how many more spheres of authority will emerge in the environmental governance of offshore oil and gas production?

7.2 Shifts in governance: how innovative is environmental governance at sea?

In the beginning of this thesis, it was argued that the protection of the marine environment presented states and other actors with extraordinary challenges, first and foremost because a large part of the marine environment is a commons and does not fall under sovereignty of a state. But also because many stakeholders have a stake in developing environmental governance initiatives and because protecting the marine environment requires transnational cooperation. It was therefore suggested that governance at sea has the potential of being a breeding ground for innovative governance arrangements with a limited role for the state. As such it would thus also be a good case for studying and gaining insight into changing governance practices. In this section, I will answer the question whether the environmental governance of shipping and offshore oil and gas production has lived up to these expectations and why (not)? What can I, based on this thesis, conclude about how innovative the governance of the marine environment really is?

The first conclusion is that one of the most unusual aspects of the environmental governance of shipping and of offshore oil and gas production is that the first spheres of authority were transnational and not national. Environmental governance on land has undergone the development from a local and national start at the end of the 1960s and early 1970s to an acceleration of transnational cooperation in the early 1990s in issue areas such as climate change and biodiversity. Instead, in the environmental governance of shipping and of offshore oil and gas production, a transnational sphere of authority emerged already in the early 1950s and 1970s respectively, even before a national sphere of authority began to take shape. Moreover, in shipping, such a national sphere of authority has never truly emerged, because the national dimension is limited to implementing IMO standards.

Still, after these transnational spheres of authority, other spheres of authority have emerged at the European level and on the national level and within the industry in the offshore oil and gas production case. In other words, the governance of the marine environment has not seen a shift from a single level, i.e. the national level, to multiple levels, but from a transnational level to national and other transnational levels. And this has indeed been different from environmental governance on land.

However, a distinction should be made between those activities that are cross-boundary and those that are not. In that sense, the cases might have given a biased picture with regard to governance at sea. Activities like shipping and offshore oil and gas production might be transboundary, similar to fishing. However, the building of wind mill farms and activities like sand and gravel extraction are not necessarily so. Those are therefore expected to be characterized by a more national way of governing. Still, the protection of the marine environment seems to provoke more transnational spheres of authority than on land. Especially in recent years, as the European Union is developing its integrated maritime policy, is developing ecosystem based approaches for the marine environment and has advocated the establishment of marine protected areas.

Second, I conclude that in both cases, the authority of the state has been stronger than expected. Especially in shipping, the authority of the state remains very important, because even though a shift in authority to multiple actors occurred, it was a shift to other state actors rather than to private actors. One of the explanations why states are more involved in governance the marine environment than expected has to do with the fact that states have high interests in marine resources such as fish and energy. This is also why the jurisdiction of states over coastal areas was extended from 3 nautical miles to 200 nautical miles through UNCLOS. A second explanation is that states have also high (economic) interests in the activities of shipping and oil and gas production. On top of that, the ship itself was already under jurisdiction of the state through the concept of flag states. That also explains why these states have always been important in developing standards for shipping. Finally, marine pollution might happen far away from the coast, but it is the experiences of coastal states with this pollution that has driven calls for environmental governance.

In line with that, the third conclusion is that the authority of private actors has been less than expected. The industry has gained authority in the environmental governance of offshore oil and gas production, but their authority is increasingly being confined by the emergence of the EU sphere of authority. In the environmental governance of shipping, the industry has been involved in a more conventional way, i.e. generating authority by providing knowledge, expertise and technologies to deal with environmental issues. In addition, environmental NGOs only play a minor role in the environmental governance of shipping and offshore oil and gas production. Their authority also remains rather conventional, meaning that they mostly have a critical voice in negotiations and try to put pressure on policy makers to adopt stringent standards. All in all, private actors have either not taken on the broader role as was suggested in the introduction of this thesis or their broader role is challenged by new developments. To put it differently, in terms of actors involved, governance at sea is more alike land-based governance practices than originally expected.

An explanation for the somewhat limited role of environmental NGOs and other civil society actors is the lack of public opinion that exists around shipping, offshore oil and gas production and other maritime activities. This lack of public opinion exists because the two maritime activities take place out at sea and are therefore not very visible to the wider public. There have been instances when these sectors aroused a lot of public interest and especially concern. But this has always been related to incidents such as the Exxon Valdez or Brent Spar and is not a structural feature.

In sum, it can be concluded that the environmental governance of the sea is more similar to land-based environmental governance than expected at the start of this thesis. There is one important difference, however, namely the traditional transnational character of the governance of the marine environment. Yet, since this transnational character was linked to state actors' cooperation, formal rules of the game and formal, governmental steering mechanisms, the governance of the marine environment has had a remarkably similar and conventional start as land-based environmental governance at least in terms of state involvement. Moreover, as was concluded in the previous chapter, the governance of the marine environment has to a large extent undergone similar shifts in governance, namely a shift to multiple levels, to multiple actors, to multiple rules and to multiple steering mechanisms.

7.3 Changes in governance and authority

This research also serves as a basis for conclusions about how and why governance practices are changing and the subsequent changing authority of the state. Against the background of the changing political landscape through globalization, shifts in governance and the changing authority of the state is subject to debate and empirical research for roughly two decades now. In this section, the insights to that debate generated by this thesis will be discussed. Since this debate is often formulated in terms of conventional versus new or innovative governance arrangements, I will reflect on that distinction first. Second, in Section 7.3.2, conclusions will be drawn about how changes in governance occur. In Section 7.3.3, conclusions on the changing authority of the state will be drawn.

7.3.1 Conventional and new spheres of authority

Although literature on governance is very diverse (see for example Pierre & Peters, 2000; Van Kersbergen & Van Waarden, 2004), an important part of the governance debate has evolved around characterizing new and innovative governance practices and to compare them with conventional governance practices (see for example Treib *et al.*, 2007). How do these practices work, who is involved, which rules and institutions guide these practices, how do they govern? And how is that different from conventional state-based governance arrangements? Based on this thesis, there are three conclusions that can be drawn about the way in which this distinction plays a role in research into governance practices.

First, one of the issues and focus of much empirical research is to generate more insight into new governance arrangements or spheres of authority. These are subsequently compared against conventional spheres of authority by using the theoretical reference point of territorialized, state-based governance with formal rules and formal law as outcome. Thus, empirical claims about new governance arrangements are compared with a theoretical construct about how conventional governance arrangements are organized and how they function. Conclusions about shifts in governance are thus based on a mix between theoretical and empirical claims.

This research shows, however, that the theoretical construct of conventional governance is not always a fair starting point for such a comparison. Yes, the 'traditional' spheres of authority in this research were very much state based, but not necessarily territorialized. In fact, in both cases the initial spheres of authority were transnational in nature. Second, and more importantly, in both spheres of authority private actors were already involved and had authority. In other words, the traditional spheres of authority were different from the theoretical picture that exists of conventional spheres of authority. What is more, based on this difference, one can argue that the activities and authority of private actors tend to be underestimated in the theoretical picture that is drawn about conventional governance. For example, private actors in the environmental governance of the two cases have always had an important role in generating compliance, because state based enforcement mechanisms were lacking. What can thus be concluded from this thesis is that the theoretical narrative of conventional governance can sometimes be questioned. Moreover, it can be concluded that making an empirical comparison gives a more grounded knowledge about changes in governance than solely relying on theoretical constructs.

Second, one of the general assumptions of much governance literature is that new spheres of authority are characterized by public-private cooperation or by no or little participation of the state. In other words, the increased private character of governance is emphasized in governance literature. In this research, such new spheres of authority indeed emerged in the case of offshore oil and gas production where the national spheres of authority included clear initiatives for public-private cooperation and the industry became a source of authority as well. However, this research also shows that new spheres of authority can be more public in nature. For example, the EU and its member states emerged as a new source of public authority. The conclusion of this thesis is therefore that new spheres of authority can consist of public-private cooperation but can also show a new form of state-based governance.

Finally, a consequence of a focus on new governance arrangements is a negligence of the way in which new governance arrangements co-exist with already existing, conventional governance arrangements. As already concluded in the previous chapter, the traditional spheres of authority continue to exist, despite several policy changes. This research has thus shown empirically what Rosenau has already argued theoretically about the development of a bifurcated world of politics. This thesis shows that newer spheres of authority are often closely linked to spheres of authority that already existed and do not always function autonomously. Rather, they should be put in the context in which they function and part of this context consists of an already existing sphere of authority.

In sum, the study of new spheres of authority has given important insights into changes in the way environmental issues are governed, and the usefulness of making a theoretical distinction between conventional and new governance arrangements should not be denied. Yet, this thesis shows some of the drawbacks in working with such a distinction and in studying only new spheres of authority. I therefore argue that new spheres of authority should be studied in their broader context. This broader context includes more empirical insight in how traditional spheres of authority functioned and how they differ from newer spheres of authority. But it also includes empirical studies into how new spheres of authority are related to already existing spheres of authority.

7.3.2 Changing governance

The objective behind studying new spheres of authority has been to generate insight into how the political landscape is changing. The concepts policy change and renewal were developed exactly for this purpose. These two concepts are also developed in an attempt to bridge the gap between the driving forces behind changes and the detailed changes within specific policy domains. Both these issues are discussed in separate theoretical domains; the driving forces behind shifts in governance are discussed within globalization literature, while the results of shifts in governance are discussed within governance literature. However, making a link between globalization as the reason why the political landscape is changing and the exact results of such changes within a certain policy domain and how they transform the governance lay-out of a certain policy domain is usually not done. In fact, I would argue that governance theory is, due to this negligence, ill-equipped to study how changes actually come about.

This thesis has made a first step in filling the link between globalization and the emergence of changes within governance practices. It was assumed that changes in governance can occur through two mechanisms; either through changes in the existing sphere of authority or through the emergence of a new sphere of authority. In the previous chapter, the conclusion was drawn that the substantial changes in the two case studies in this thesis were driven by external developments, most of them with the result of the emergence of a new sphere or source of authority. In other words, substantial changes in environmental governance are most likely to be related to developments that occur outside of the existing sphere(s) of authority and not as a result of changes within the existing spheres of authority.

The first question that then arises is whether these developments are indeed evidence of globalization. Part of globalization is the globalization of governance practices. Looking at the developments within the environmental governance of shipping and offshore oil and gas production from this angle, one could argue that the policy changes shows signs of the globalization of governance practices. Even though the traditional spheres of authority were already transnational, the adoption of UNCLOS and the Europeanization of the policy domain of the marine environment shows a further extension of the globalization of governance practices in protecting the marine environment. Yet, it should be noted that this thesis also analysed the emergence of national spheres of authority as one of the policy changes in the environmental governance of offshore oil and gas production. What does that say about globalization as a driving force for policy changes in governance? This is a more difficult question to answer. The emergence might be a reaction to the globalization of governance practices, a phenomenon know as glocalization. Maybe the emergence of national spheres of authority was a reaction to authority moving away from the national level to the regional level and to the private sector?

In addition, the increased participation of private actors in governance practices is also considered to be part of globalization. This thesis shows mixed results with regard to the increased authority for private actors. In both the environmental governance of shipping and of offshore oil and gas production, the authority of private actors has been undermined by the emergence of the EU sphere of authority. One of the lessons for globalization literature from this thesis is therefore that the globalization of governance practices (i.e. the Europeanization of the marine environment) might sometimes conflict with the globalization of private authority in governance practices.

Another new piece contributed by this thesis for the puzzle of where change in governance comes from is that not only new spheres of authority emerged, but existing spheres of authority went through a process of renewal as a result of policy changes as well. The newly emerged spheres of authority also affects existing sphere(s) of authority because of overlap and interaction between new and existing spheres of authority. The first conclusion is therefore that studying only new spheres of authority does not suffice when one wants to understand and explain how changes in governance come about. It is important to focus on how new spheres of authority cause change in existing spheres of authority to fully understand changes in governance in a certain policy domain. This confirms the conclusion that understanding processes of change in governance requires a study of new spheres of authority in their context. This context consists of both the development that causes a policy change as well as the effect the policy change has on the existing sphere(s) of authority.

7.3.3 *Changing authority of the state*

Central to the debate in governance is not only how and why governance is changing, but also what implications these shifts in governance have for the authority of the state. Implicit in these debates is the assumption that because new sites of governance emerge and private actors become increasingly involved in governing, the state is losing authority. This thesis contributes to this debate in several ways.

Generally speaking, the two case studies indeed show a diffusion of authority to different actors as governance literature suggests. The first conclusion that can be drawn, however, is that this diffusion of authority is not a zero-sum game with less authority for states and more for non-state actors. Rather, states can remain authoritative, but are increasingly sharing authority with other actors. Related to that, I conclude that these other actors do not necessarily have to be private actors, something that is contrary to what is suggested in governance literature. After all, this research shows that new (groups of) states can gain authority and that environmental NGOs or the industry are not always the ones gaining authority.

There are also two conclusions that can be drawn about why the authority of the state is changing. The first is that both the renewal of an existing sphere of authority and the emergence of a new sphere of authority are sources of change. More particularly, it can be concluded that the diffusion of authority to non-state actors is more likely in new spheres of authority, where the breeding ground for innovation is better than in existing spheres of authority. In other words, in existing spheres of authority, a change will most probably lead to a mix of conventional and new ways in which the state generates authority, while in new spheres of authority there is a higher chance that the authority of actors is newly defined. It should be noted that this conclusion is based on this thesis only and that further research should be carried out to see whether this conclusion is also valid for other policy domains.

The second main lesson learned from this research is thus that the role of the state is diversifying rather than changing from its conventional role to a new role. It is diversifying, because a state is increasingly involved in different spheres of authority and has to deal with competing or overlapping spheres and sources of authority that emerge. Moreover, the diffusion of authority also means that both the state and the industry have to find ways to continue to communicate and coordinate the demands that are put on the state and the industry by different spheres of authority.

Finally, the argument made above that changes in governance and new spheres of authority should always be studied in their context, also holds for studying changes in the authority of the state. In this research, the 'traditional' authority of the state is not an isolated actor that takes decisions for the target group it regulates. States have always had to do with stakeholders and their interests, especially the target groups of the policies developed. The important question is whether the interaction between states and industry has changed and whether this is related to diffusion of authority or not. In addition, the authority of the state shows traditional elements also in renewed and new sphere of authority. Thus, in line with the conclusions above, this research shows that it is important to have an empirical account rather than only a theoretical construct of the traditional authority of the state if one wants to draw conclusions about changes in governance and the authority of the state.

7.4 Reflection on research and the conceptual framework

An important issue in reflecting on this research is that it was sometimes difficult to match the objective of analysing processes of change in governance from a long-term perspective with the objective of studying the implications for the authority of the state in more detail. Studying changes in authority of actors requires in-depth understanding of the interaction between actors in the sphere of authority and the influence actors have on the results of the sphere of authority. Finding a balance between a long-term focus and this type of in-depth research has been difficult. Overall, the long-term focus has been somewhat more dominant than going more in-depth into the exact substance, organization and results of the different spheres of authority. Even though this has been done deliberately, the decision to do so was also influenced by constraints with regard to data resources and time available for this research, especially in the case of offshore oil and gas production which has multiple spheres of authority. A recommendation for future research is therefore to follow-up the research done in this thesis with in-depth research into more specific policy issues within the marine environment, such as air pollution by shipping or chemicals discharged by the offshore industry.

This thesis has also sought to make a contribution to the development of new concepts to understand and analyse changes in governance and in the authority of actors. The concept sphere of authority has been proposed as a concept to study various properties of governance practices so that older and newer spheres of authority can be compared with regard to their substance, organization and results. The research in this thesis has proven that this concept is broad enough to cover various types of governance arrangements, including conventional ones.

It has also proven to be a concept that can be used to analyse the dynamics within a sphere of authority and the resulting authority of actors in developing steering mechanisms and generating compliance. In fact, even though this thesis has explicitly sought to analyse the authority of the state, it should be noted that this model also generates insight in the changing authority of other actors. It is a conceptual model that allows studies into changing governance practices and that can be used to reflect on the authority of all actors involved in the sphere of authority.

Still, it should also be noted that because of the long-term perspective in this thesis, the organization, substance and results of the spheres of authority in the environmental governance of shipping and of offshore oil and gas production have not always been studied as in-depth as would have been desirable. The true merits of this conceptual model in linking the dynamics within a sphere of authority with the resulting authority of actors in developing steering mechanisms and generating compliance can therefore only be further assessed when in-depth studies are done.

The conceptual model of this thesis also generates insight in processes of change in governance through the concepts policy change and renewal of a sphere of authority. Long-term changes can be analysed through the concept policy change, while more detailed changes within spheres of authority can be assessed through the renewal of the organization, substance and results of governance practices. This combination also specifically allows for a study into the interaction between the emergence of new sphere of authority and change in existing spheres of authority.

However, this research has also shown that it remains difficult to pinpoint to when something can be regarded as a policy change or as the renewal of a sphere of authority and when not. It was argued that this is an empirical question and is based on the interpretation of the researcher. Although I still subscribe this claim, further research into how changes in governance come about is necessary to enhance our understanding and to refine the concepts used in this thesis.

Something that can also be learned from the application of the conceptual model in this thesis is that in some cases steering mechanisms also served as the rules of the game within a spheres of authority. For example, the Dutch Environmental Covenant for the offshore industry is a steering mechanism because it consists of a set of objectives and standards that should be reached. However, the Environmental Covenant also describes procedures for the development of Company Environmental Plans. As such the Environmental Covenant is also a source for rules of the game. In a similar fashion, the MARPOL Convention is a steering mechanism with its environmental targets and equipment standards, but it also contains some rules of the game regarding rights for port states to conduct Port State Control. This can be rather confusing, but apparently such kinds of steering mechanisms can serve two purposes: to set objectives and to change behaviour or operations, or to guide procedures and to assign actors specific rights and obligations with regard to decision making and implementation.

Finally, although this thesis has tried to contribute to making a link between globalization literature which focuses on long-term changes in society and the more detailed governance literature on the changing political landscape, it has proven to be difficult to bridge this gap. The idea of policy change has been one step in that direction, however, the link between the causes of policy changes and globalization is not covered by the conceptual framework of this thesis nor in the analysis done in the empirical chapters. This therefore remains a challenge for future theorizing and research.

A final issue in the reflection on the research done in this thesis and the conceptual framework is the generalization of this research. In Chapter 3, it was argued that generalization from the two maritime cases to other maritime cases and from the North Sea to other seas is not possible, because of the specific context of the cases studied in this thesis. Yet, one issue is more generally experienced, at least within the European context. The increased importance of EU maritime policy will be similar for other maritime activities and seas in Europe. Seas like the Baltic Sea and the Mediterranean Sea are similar to the North Sea in the sense that they are semi-enclosed, have multiple activities taking place, know a regional Convention like the OSPAR Convention and are getting increased attention of EU legislation. This means that some of the dynamics between the already existing regional spheres of authority and the EU sphere of authority in those areas will be similar to the dynamics between the OSPAR and EU spheres of authority discussed in this thesis.

Yet, generalizing results of this thesis to other maritime activities and regions outside the EU is much more difficult. Shipping and offshore oil and gas production take place on a global and regional scale. Only the activity of fisheries is also taking place on a transnational scale. Yet, fisheries governance cannot be compared to the environmental governance of shipping or of offshore oil and gas production, because the authority of the EU is of a different nature. Fisheries governance within Europe evolves around the EU since the 1970s. The realm of fisheries is even one of the few policy areas in which the EU has supranational powers. The long existing authority and the influence of the EU in fisheries differs therefore substantially from its emerging role in shipping and offshore oil and gas production. Contrary to shipping, offshore oil and gas production and fisheries, other maritime activities lack this transnational character. For example, even though the development of offshore wind mill farms is spreading across the North Sea, the activities itself are nationally focused and governed.

Besides the limited generalization of the two maritime cases and the North Sea to other activities and regions, there is also the generalization of the conceptual framework to other maritime activities, marine regions and the environmental policy domain. Since the conceptual framework developed in this thesis is not dependent on specific governance levels, actors, rules, steering or compliance mechanisms, the conceptual framework has the potential of application across a broad variety of policy domains. While the application of the concepts sphere of authority and policy change onto other maritime activities, marine regions and the environmental policy domain seems most logical, I do not want to exclude other public policy domain such as education, health, agriculture or food. Yet, further research is needed to prove the true generalizability of these concepts.

7.5 Epilogue: future outlook for environmental governance at sea

There is no point in denying that governing the marine environment is becoming increasingly important. Activities are moving from land to sea and pressure on the oceans as a source of resources is increasing as well. In contrast to this thesis, which has explored the past development of governance practices aimed to protect the marine environment, this epilogue will take a look at the future. What can be expected about the future of governance for the marine environment and in particular with regard to shipping and offshore oil and gas production?

In shipping, the main question is whether the IMO can continue to be the only body developing standards with this increased attention for maritime activities and the marine environment. I indeed expect more challenges for IMO from the EU, the shipping industry and other international regimes. The EU is developing integrated solutions for the protection of the marine environment which will also affect the sector of shipping. In addition, the EU is currently focusing on air emissions and is not happy with the progress made within IMO with regard to the reduction of greenhouse gases. Moreover, the post-2012 negotiations on climate change are underway and with the lack of activities within IMO this might well result in the inclusion of shipping in the post-

2012 regime for climate change. Finally, the industry is roaring its tail because it is realizing that sustainability and greening their operations can also have market potential. At the same time, cargo owners hiring ships to transport goods are also focusing increasingly on setting environmental requirements for the ships they want to hire. These are all potential spheres of authority that challenge IMO and it will depend on IMO and how effective the IMO can overcome these challenges, whether it will remain the single standards setting body in shipping. My expectation is that they will not be able to combat all these challenges at the same time and at the speed that is currently required. One of the main reasons is because IMO has a history in being reactive instead of proactive and in being slow in negotiating and implementing standards.

The offshore oil and gas industry faces a different future. Offshore oil and gas production in the North Sea is beyond its peak. For the coming decades, however, oil and gas is needed and will be produced in the North Sea. However, with North Sea oil and gas resources being over their peak, I expect that the pressure on greening the production of oil and gas in that region will lessen. This in contrast to the Barents Sea, where new gas fields are exploited in harsh conditions and where pressure to innovate and have zero emissions has already become very important. For the time that offshore oil and gas in the North Sea remains an important activity, I expect an increasing importance of the EU for the development of standards and policies. The EU is developing the integrated marine policy and is applying the habitats directive onto European Seas. Since platforms are static and knowledge on discharges and emissions is available, they are an easy target in these policies. Moreover, I expect that a close relationship between the EU and OSPAR will develop for the implementation of these European policies. Together with the end of the Dutch environmental covenant and an implemented zero discharge policy, I think the national level will lose some of its importance, while the regional level, and especially the EU, will become more important in the environmental governance of offshore oil and gas production.

Going beyond shipping and offshore oil and gas production, the maritime activities at sea, which have always been governed by sector specific spheres of authority, will get closer ties, because of the establishment of marine protected areas, the call for ecosystem based governance and because of the integrated maritime policy of the EU. What is missing on land, i.e. an overall set of objectives for protecting and governing the environment, is currently under development in the European Union for the marine environment. This will present states, the industry and civil society actors with extraordinary challenges. They will have to compete and share authority with actors from other sectors during the development and implementation of these kinds of policies. It is these developments that, despite evidence from this thesis, make me expect that the marine environment will, at least in Europe, become a breeding ground for changing governance practices.

References

Arts, B., & Leroy, P. (2006). *Institutional Dynamics in Environmental Governance*. Dordrecht: Springer.

Arts, B., Leroy, P., & Van Tatenhove, J. (2006). Political Modernisation and Policy Arrangements: A Framework for Understanding Environmental Policy Change. *Public Organization Review, 6*(2), 93-106.

Arts, B., & Van Tatenhove, J. (2004). Policy and Power: A Conceptual Framework between the 'old' and 'new' Policy Idioms. *Policy Sciences, 37*(3-4), 339-356.

Arts, B., & Van Tatenhove, J. (2006). Political Modernisation. In B. Arts & P. Leroy (Eds.), *Institutional Dynamics in Environmental Governance* (pp. 21-43). Dordrecht: Springer.

Arts, B., Van Tatenhove, J., & Leroy, P. (2000). Policy Arrangements. In J. Van Tatenhove, B. Arts & P. Leroy (Eds.), *Political Modernisation and the Environment; The Renewal of Environmental Policy Arrangements* (pp. 53-69). Dordrecht: Kluwer Academic Publishers.

Barrett, B., & Howells, R. (1990). The Offshore Petroleum Industry and Protection of the Marine Environment. *Journal of Environmental Law, 2*(1), 23-64.

Barton, J. (1999). Flags of Convenience: Geoeconomics and Regulatory Minimisation. *Tijdschrift voor Economische en Sociale Geografie, 90*(2), 142-155.

Beck, U. (1994). The Reinvention of Politics: Towards a Theory of Reflexive Modernization. In U. Beck, A. Giddens & S. Lash (Eds.), *Reflexive Modernization; Politics, Tradition and Aesthetics in the Modern Social Order* (pp. 1-55). Cambridge: Polity Press.

Beck, U. (2005). *Power in the Global Age: a new Global Political Economy*. Cambridge: Polity Press.

Blanco-Bazán, A. (2004). IMO - Historical Highlights in the Life of a UN Agency. *Journal of the History of International Law, 6*(2), 259-283.

Boehmer-Christiansen, S. (1984). Marine pollution control in Europe. Regional approaches, 1972-80. *Marine Policy, 8*(1), 44-55.

Bruyninckx, H., Spaargaren, G., & Mol, A.P.J. (2006). Introduction: Governing Environmental Flows in Global Modernity. In F. Buttel, G. Spaargaren & A.P.J. Mol (Eds.), *Governing environmental Flows in Global Modernity* (pp. 1-31). Cambridge: MIT Press.

Burnham, P., Lutz, K.G., Grant, W., & Layton-Henry, Z. (2008). *Research Methods in Politics* (2nd edition). New York: Palgrave Macmillan.

Bus, M.F. (1993). *Het Rechtsregime met betrekking tot Milieubescherming bij Mijnbouwactiviteiten op het Nederlands Continentaal Plat*. Utrecht: Netherlands Institute for the Law of the Sea.

Champ, M.A. (2000). A Review of Organotin Regulatory Strategies, pending Actions, related Costs and Benefits. *The Science of the Total Environment, 258*(1-2), 21-71.

De Jong, J.J., Weeda, E., Westerwoudt, T., & Correljé, A. (2005). *Dertig Jaar Nederlands Energiebeleid; Van Bonzen, Polders en Markten naar Brussel zonder Koolstof*. Den Haag: Clingedeal International Energy Programme.

De Jong, J.W., & Henriquez, L.R. (2000). *A Mechanism to Reduce the Environmental Impact together*. Paper presented at the SPE International Conference on Health, Safety and the Environment in Oil and Gas Exploration and Production.

De La Fayette, L. (1999). The OSPAR Convention Comes into Force: Continuity and Progress. *The International Journal of Marine and Coastal Law, 14*(2), 247-297.

De La Fayette, L. (2001). The Marine Environment Protection Committee: The Conjunction of the Law of the Sea and International Environmental Law. *The International Journal of Marine and Coastal Law, 16*(2), 155-238.

Department of Trade and Industry (2005). *Explanatory Memorandum to the Offshore Petroleum Activities (Oil Pollution Prevention and Control) Regulations 2005*. London.

DeSombre, E.R. (2006). *Flagging Standard; Globalization and Environmental, Safety and Labor Regulations at Sea*. Cambridge (Massachusetts)/London: MIT Press.

Dingwerth, K., & Pattberg, P. (2006). Global Governance as a Perspective on World Politics. *Global Governance, 12*(2), 185-203.

Ducrotoy, J.-P., Elliot, M., & De Jonge, V.N. (2000). The North Sea. *Marine Pollution Bulletin, 41*, 5-23.

E & P Forum (1994). *Guidelines for the Development and Application of Health, Safety and Environmental Management Systems* (No. 6.36/210). London.

Environmental & Safety Consultancy (1992). *Decisions of the Paris Commission with regard to the Offshore Industry within the Netherlands, United Kingdom and Norway*. Den Helder.

European Commission. (2005). Press Release - Maritime Safety and Port State Control: European Commission requires proper Implementation of the Rules. Retrieved 19 June 2009, from http://europa.eu/rapid/pressReleasesAction.do?reference=IP/05/1631&format=HTML&aged=1&language=EN&guiLanguage=en.

European Commission. (2006). Press Release - Port State Control: Commission sends a reasoned Opinion to Portugal and brings Malta to the Court of Justice. Retrieved 19 June 2009, from http://europa.eu/rapid/pressReleasesAction.do?reference=IP/06/883&format=HTML&aged=1&language=EN&guiLanguage=en.

European Commission (2007). *Guidelines for the Establishment of the Natura 2000 Network in the Marine Environment; Application of the Habitats and Birds Directives*. Brussels.

European Commission. (2009). 3rd Maritime Safety Package - Port State Control. Retrieved 17 June 2009, from http://ec.europa.eu/transport/maritime/safety/doc/2009_03_11_package_3/fiche03_en.pdf.

Farthing, B. (1993). *International Shipping; An Introduction to the Policies, Politics and Institutions of the Maritime World* (2nd edition). London: Lloyd's of London Press.

Fitzmaurice, V. (1978). Legal Control of Pollution from North Sea Petroleum Development. *Marine Pollution Bulletin, 9*(6), 153-156.

Flare Transfer Pilot Trading Scheme Steering Committee (2002). *FTPTS Background*. London.

Flare Transfer Pilot Trading Scheme Steering Committee (2006). *Flare Transfer Pilot Trading Scheme; Annual Report 2005*. London.

FO-Industrie (2006). *Handreiking BMP-4 Olie- en Gaswinningsindustrie; Een Handreiking bij het Opstellen van het BMP-4 van de Olie- en Gaswinnende Industrie*. Den Haag.

FO-Industrie (2008). *Uitvoering Intentieverklaring Olie- en Gaswinningsindustrie, Sommatierapport BMP-4*. Den Haag.

FO-Industrie, & PricewaterhouseCoopers N.V. (2001). *Evaluatie Intentieverklaring Milieubeleid Olie- en Gaswinningsindustrie; Naar Randvoorwaarden voor een Goede Doorstart*. Den Haag/Utrecht.

Frank, V. (2005). Consequences of the Prestige Sinking for European and International Law. *The International Journal of Marine and Coastal Law, 20*(1), 1-64.

Furger (1997). Accountability and Systems of Self-Governance: The Case of the Maritime Industry. *Law & policy, 19*(4), 445-476.

Gilpin, R. (2002). A Realist Perspective on International Governance. In D. Held & A. McGrew (Eds.), *Governing Globalization; Power, Authority and Global Governance* (pp. 237-248). Malden: Polity Press and Blackwell Publishing Ltd.

Gray, J.S., Bakke, T., Beck, H.J., & Nilssen, I. (1999). Managing the Environmental Effects of the Norwegian Oil and Gas Industry: From Conflict to Consensus. *Marine Pollution Bulletin, 38*(7), 525-530.

Greenaward. (2009a). Benefits. Retrieved 25 June 2009, from http://www.greenaward.org/defaulthome.htm.

Greenaward. (2009b). Which Shipowners and Vessels? Retrieved 25 June 2009, from http://www.greenaward.org/defaulthome.htm.

Hajer, M. (2003). Policy without Polity? Policy Analysis and the Institutional Void. *Policy Sciences, 36*(2), 175-195.

Halpern, B.S., Walbridge, S., Selkoe, K.A., Kappel, C.V., Micheli, F., D'Agrosa, C., *et al.* (2008). A Global Map of Human Impact on Marine Ecosystems. *Science, 319*, 948-952.

Held, D., & McGrew, A. (2002). *Governing Globalization; Power, Authority and Global Governance*. Malden: Polity Press and Blackwell Publishing Ltd.

Held, D., McGrew, A., Goldblatt, D., & Perraton, J. (1999). *Global Transformations; Politics, Economics and Culture*. Stanford: Stanford University Press.

Hey, E. (1993). The 1992 Paris Convention for the Protection of the Marine Environment of the North-East Atlantic: A Critical Analysis. *The International Journal of Marine and Coastal Law, 8*(1), 1-76.

Hirst, P., & Thompson, G. (1996). *Globalization in Question; The International Economy and the Possibilities of Governance*. Cambridge (UK): Polity Press.

Hooghe, L., & Marks, G. (2001). *Multi-level Governance and European Integration*. Lanham/Oxford: Rowman & Littlefield Publishers.

Hoppe, H. (2000). *Port State Control - an update on IMO's Work*. London: International Maritime Organization.

Hornby, A.S. (1995). *Oxford Advanced Learner's Dictionary of Current English*. Oxford: Oxford University Press.

IJlstra, T. (1990). Laws of the Sea; Air Pollution from Shipping. *Marine Pollution Bulletin, 21*(7), 319-320.

IncidentNews. (1977). Ekofisk Bravo Oil Field. Retrieved 21 August 2008, from http://www.incidentnews.gov/incident/6237.

International Chamber of Shipping, & International Shipping Federation (2006). *Annual Review 2005*. London.

International Maritime Organization (1998a). *Focus on IMO; IMO 1948-1998: a Process of Change*. London.

International Maritime Organization (1998b). *Focus on IMO; MARPOL - 25 years*. London.

International Maritime Organization. (2001). Sub-Committee on Flag State Implementation - 9th session: 19-23 February 2001. Retrieved 3 July 2008, from http://www.imo.org/.

International Maritime Organization. (2002). International Convention for the Prevention of Pollution from Ships, 1973, as modified by the Protocol of 1978 relating thereto (MARPOL 73/78). Retrieved 29 January 2008, from http://www.imo.org/.

International Maritime Organization. (2003). Sub-Committee on Flag State Implementation - 11th session: 7-11 April 2003. Retrieved 3 July 2008, from http://www.imo.org/.

International Maritime Organization. (2008). IMO: 50 years, 50 treaties. Retrieved 18 March 2008, from http://www.imo.org/Newsroom/mainfram.asp?topic_id=1709&doc_id=9076.

Intertanko. (2001). History - Overview. Retrieved 7 April 2008, from http://www.intertanko.com/templates/Page.aspx?id=1069.

Jagota, S.P. (2000). Developments in the Law of the Sea between 1970 and 1998: A Historical Perspective. *Journal of the History of International Law, 2*(1), 91-119.

Joint Nature Conservation Committee. (2009). SACs in UK Offshore Waters. Retrieved 20 August 2009, from http://www.jncc.gov.uk/protectedsites/sacselection/SAC_list.asp?Country=OF.

Karman, C.C., & Tamis, J.E. (2000). *Syllabus International Course on Environmental Practices in Offshore E&P Activities*. Apeldoorn: TNO-MEP.

Keohane, R.O. (2002). *Power and Governance in a Partially Globalized World*. London: Routledge.

Keselj, T. (1999). Port State Jurisdiction in Respect of Pollution from Ships: The 1982 United Nations Convention on the Law of the Sea and the Memorandum of Understanding. *Ocean Development and International Law, 30*(2), 127-160.

Kjaer, A.M. (2004). *Governance*. Cambridge/Malden: Polity Press.

Klijn, E.-H., & Koppenjan, J.F.M. (1997). Beleidsnetwerken als Theoretische Benadering: Een Tussenbalans. *Beleidswetenschap*(2), 143-167.

Krosby, H.P. (1976). Oil and the Environment: Norway's Enlightened Policy. *Scandinavian Review, 64*(4), 38-44.

Kütting, G., & Gauci, G. (1996). International Environmental Policy on Air Pollution from Ships. *Environmental Politics, 5*(2), 345-352.

Leroy, P., & Arts, B. (2006). Institutional Dynamics in Environmental Governance. In B. Arts & P. Leroy (Eds.), *Institutional Dynamics in Environmental Governance* (pp. 1-19). Dordrecht: Springer.

Liefferink, D. (2006). The Dynamics of Policy Arrangements: Turning Round the Tetrahedron. In B. Arts & P. Leroy (Eds.), *Institutional Dynamics in Environmental Governance* (pp. 45-68). Dordrecht: Springer.

M'Gonigle, R.M., & Zacher, M.W. (1979). *Pollution, Politics, and International Law; Tankers at Sea*. Berkeley/Los Angelos: University of California Press.

Marbry, L. (2008). Case Study in Social Research. In P. Alasuutari, L. Bickman & J. Brannen (Eds.), *The SAGE Handbook of Social Research Methods*. London: SAGE Publications Ltd.

Marine Environment Protection Committee (1984). *Report of the Marine Environment Protection Committee on its Ninthteenth Session*. London.

Marine Environment Protection Committee (1988). *List of Documents issued in connection with the Twenty-sixth Session of the Marine Environment Protection Committee*. London.

Marine Environment Protection Committee (2003). *Report of the Marine Environment Protection Committee on its Forty-ninth Session*. London.

Marine Environment Protection Committee (2006). *Report of the Marine Environment Protection Committee on its Fifty-fifth Session*. London.

Marine Environment Protection Committee (2007). *Report of the Marine Environment Protection Committee on its Fifty-sixth Session*. London.

Marine Environment Protection Committee (2008). *Report of the Marine Environment Protection Committee on its Fifty-seventh Session*. London.

Marine Pollution Bulletin (1975). Flags of Convenience. *Marine Pollution Bulletin, 6*(4), 49-51.

Marks, G., Hooghe, L., & Blank, K. (1996). European Integration since the 1980s. State-Centric versus Multi-Level Governance. *Journal of Common Market Studies 34*, 343-378.

Marthinsen, I., & Sørgård, T. (2002). *Zero Discharge Philosophy: A Joint Project between Norwegian Authorities and Industry*. Paper presented at the SPE International Conference on Health, Safety and the Environment in Oil and Gas Exploration and Production.

McConnell, M.L. (2002). *GloBallast Legislative Review; Final Report* (No. 1). London: Global Ballast Water Management Programme; International Maritime Organization.

Ministerie van Economische Zaken (2003). *Van Mijnwetten naar Mijnbouwwet*. Den Haag.

Ministerie van Economische Zaken, & Ministerie van Volkshuisvesting; Ruimtelijke Ordening en Milieubeheer (1996). *Omgaan met nieuwe Milieubeleid in Doelgroepenbeleid Industrie*. Den Haag.

Ministerie van Volkshuisvesting; Ruimtelijke Ordening en Milieubeheer, Ministerie van Economische Zaken, Ministerie van Landbouw en Visserij, & Ministerie van Verkeer en Waterstaat (1989). *Nationaal Milieubeleidsplan; Kiezen of Verliezen*. Den Haag.

Ministry of Environment (1996-97). *Report to the Storting No. 58 Environmental Policy for a Sustainable Development; Joint Efforts for the Future*. Oslo

Mitchell, R.B. (1994). *Intentional Oil Pollution at Sea; Environmental Policy and Treaty Compliance*. Cambridge: MIT Press.

Molenaar, E.J. (1996). The EC Directive on Port State Control in Context. *The International Journal of Marine and Coastal Law, 11*(2), 241-288.

Molenaar, E.J. (2007). Port State Jurisdiction: Toward Comprehensive, Mandatory and Global Coverage. *Ocean Development and International Law, 38*(1&2), 225-257.

Nelsen, B.F. (1991). *The State Offshore; Petroleum, Politics, and State Intervention on the British and Norwegian Continental Shelves*. New York: Praeger.

Netherlands Oil and Gas Exploration and Production Association (2002). *Milieujaarrapportage 2001*. Den Haag: FO-Industrie.

Nilssen, I., & Øren, H.M. (2003). Zero Discharges to Sea from Petroleum Activity on the Norwegian Continental Shelf. *Business Briefing: Exploration & Production: the Oil & Gas Review, 2*, 51-58.

Ninaber, E. (1997). Monitoring Policy and Law; Marpol Annex VI A licence to pollute. *North Sea Monitor*, 7-9.

Nollkaemper, A., & Hey, E. (1995). Implementation of the LOS Convention at Regional Level: European Community Competence in Regulating Safety and Environmental Aspects of Shipping. *The International Journal of Marine and Coastal Law, 10*(2), 281-300.

North Sea Directorate (2002). *Ballast Water Management in the North-East Atlantic; Report to aid decision making on Ballast Water in OSPAR BDC*. The Hague.

North Sea Foundation. (no date). Zero Emission Ship. Retrieved 25 June 2009, from http://www.cleanship.info/the%20cleanship%20concept/zero%20emission%20ship.htm.

North Sea Ministerial Conference (1984). *Bremen Declaration; Declaration of the International Conference on the Protection of the North Sea*. Bremen.

North Sea Ministerial Conference (1987). *London Declaration; Second International Conference on the Protection of the North Sea, London, 24-25 November 1987*. London.

North Sea Ministerial Conference (1990). *The Hague Declaration; Ministerial Declaration of the Third International Conference on the Protection of the North Sea*. The Hague.

North Sea Ministerial Conference (1995). *Esbjerg Declaration; Ministerial Declaration of the Fourth International Conference on the Protection of the North Sea*. Esbjerg

North Sea Ministerial Conference (2002). *Bergen Declaration; Fifth International Conference on the Protection of the North Sea*. Bergen.

North Sea Ministerial Conference (2006). *Göteburg Declaration; North Sea Ministerial Meeting on the Environmental Impact of Shipping and Fisheries*. Göteborg.

Norwegian Oil Industry Association (2007). *Environmental Report 2006*. Stavanger.

Norwegian Petroleum Directorate (2000). Facing up to green realities. *Norwegian Petroleum Diary, 1*, 3-5.

Oil & Gas Industry Task Force (1999). *The Oil & Gas Industry Task Force Report; A Template for Change*. Aberdeen.

Oil & Gas UK (2007). *Oil & Gas UK Sustainable Development Report*. London.

Pallis, A.A. (2006). Institutional Dynamism in EU Policy-Making: The Evolution of the EU Maritime Safety Policy. *Journal of European Integration, 28*(2), 137-157.

Paris Memorandum of Understanding on Port State Control (1998). *1997 Annual Report*. The Hague.

Paris Memorandum of Understanding on Port State Control (2005). *Annual Report 2004; Changing course*. The Hague.

Peet, G. (1991). Laws of the Sea; Ship to Air Pollution. *Marine Pollution Bulletin, 22*(7), 321-322.

Peet, G. (1994). The Role of (Environmental) Non-governmental Organizations at the Marine Environment Protection Committee (MEPC) of the International Maritime Organization (IMO), and at the London Dumping Convention (LDC). *Ocean & Coastal Management, 22*(1), 3-18.

Peet, G. (2008). Tanker Design: Recent Developments from an Environmental Perspective. In S.T. Orszulik (Ed.), *Environmental Technology in the Oil Industry*. Dordrecht/London: Springer.

Petroleum Safety Authority Norway. (2008). The Continental Shelf. Retrieved 02 September 2008, from http://www. ptil.no/regulations/the-continental-shelf-article4246-87.html.

Pierre, J. (2000). Introduction: Understanding Governance. In J. Pierre (Ed.), *Debating Governance; Authority, Steering, and Democracy* (pp. 1-10). Oxford: Oxford University Press.

Pierre, J., & Peters, B.G. (2000). *Governance, Politics and the State*. London: MacMillan Press LtD.

Prakash, A., & Hart, J.A. (1999). *Globalization and Governance*. London: Routledge.

Putnam, R.D. (1988). Diplomacy and Domestic Politics: The Logic of Two-Level Games. *International Organization, 42*(3), 427-460.

Read, A.D., & Blackman, R.A.A. (1980). Oily Water Discharges from Offshore North Sea Installations: A perspective. *Marine Pollution Bulletin, 11*(2), 44-47.

Rhodes, R.A.W. (2000). Governance and the Public Administration. In J. Pierre (Ed.), *Debating Governance; Authority, Steering, and Democracy* Oxford: Oxford University Press.

Roe, M. (2009). Multi-level and Polycentric Governance: Effective policymaking for Shipping. *Maritime Policy & Management, 36*(1), 39-56.

Rosenau, J.N. (1992). The Relocation of Authority in a Shrinking World. *Comparative Politics, 24*(3), 253-272.

Rosenau, J.N. (1997). *Along the Domestic-Foreign Frontier; Exploring Governance in a Turbulent World*. Cambridge: Cambridge University Press.

Rosenau, J.N. (1999). Toward an Ontology for Global Governance. In M. Hewson & T.J. Sinclair (Eds.), *Approaches to Global Governance Theory* (pp. 287-301). Albany: State University of New York Press.

Rosenau, J.N. (2000). Change, Complexity, and Governance in Globalizing Space. In J. Pierre (Ed.), *Debating Governance; Authority, Steering, and Democracy* (pp. 167-200). Oxford: Oxford University Press.

Rosenau, J.N. (2002). Governance in a New Global Order. In D. Held & A. McGrew (Eds.), *Governing Globalization; Power, Authority and Global Governance* (pp. 70-86). Malden: Polity Press and Blackwell Publishing Ltd.

Rosenau, J.N. (2003a). *Distant Proximities: Dynamics beyond Globalization*. Princeton/Oxford: Princeton University Press.

Rosenau, J.N. (2003b). Globalization and Governance: Bleak Prospects for Sustainability. *International Politics and Society* (3), 11-29.

Rosenau, J.N. (2006). *The Study of World Politics: Globalization and Governance*. London/New York: Routledge.

Rowan-Robinson, J. (2000). Environmental Regulation of the UK Offshore Oil and Gas Industry. *Journal of Energy & Natural Resources Law, 18*(3), 267-283.

Salvarani, R. (1996). The EC Directive on Port State Control: A Policy Statement. *The International Journal of Marine and Coastal Law, 11*(2), 225-231.

Scholte, J.A. (2000). *Globalization; a critical introduction*. New York: Palgrave.

Secretariat of the Oslo and Paris Commissions (1995). *Summary Record Working Group on Sea-Based Activities; Oslo 9-13 January 1995*. London.

Secretariat of the Oslo and Paris Commissions (1996). *Summary Record Working Group on Sea-Based Activities; Aberdeen 12-16 February 1996*. London.

Secretariat of the Oslo and Paris Commissions (1997). *Summary Record Working Group on Sea-Based Activities;*
    Biarritz 17-21 February 1997. London.

Secretariat of the Oslo and Paris Commissions (2000). *Review of the SEBA Working Procedures; Working Group on*
    Sea-based Activities; Amsterdam 14-18 February 2000. London.

Secretariat of the OSPAR Commission (1998). *Summary Record Meeting of the OSPAR Commission; Kingston Upon*
    Hull 21-24 June 1999. London.

Secretariat of the OSPAR Commission (2001). *Cooperation between OSPAR and the European Community; Meeting*
    of the Offshore Industry Committee (OIC); Oslo 13-16 February 2001. London.

Secretariat of the OSPAR Commission (2003). *Summary Record Meeting of the Offshore Industry Committee (OIC);*
    London 10-14 March 2003. London.

Secretariat of the OSPAR Commission (2005). *Summary Record Meeting of the Offshore Industry Committee (OIC);*
    Edinburgh 14-18 March 2005. London.

Secretariat of the OSPAR Commission (2007). *Summary Record Meeting of the Offshore Industry Committee (OIC);*
    Paris 12-16 March 2007. London.

Secretariat of the OSPAR Commission (2008). *Summary Record Meeting of the Offshore Industry Committee (OIC);*
    Bonn 10-14 March 2008. London.

Secretariat of the Paris Commission (1984). *Summary Record Eighth Meeting of the Working Group on Oil Pollution;*
    7-9 February 1984. London.

Secretariat of the Paris Commission (1986). *Tenth Meeting of the Working Group on Oil Pollution; Stockholm 4-6*
    February 1986: Summary Record of the Meeting with E&P Forum. London.

Secretariat of the Paris Commission (1987). *Summary Record Eleventh Meeting of the Working Group on Oil Pollution;*
    The Hague 3-5 February 1987. London.

Secretariat of the Paris Commission (1988). *Summary Record Twelfth Meeting of the Working Group on Oil Pollution;*
    Oslo 9-11 February 1988. London.

Secretariat of the Paris Commission (1990). *Summary Record Fourteenth Meeting of the Working Group on Oil*
    Pollution; Copenhagen 6-9 February 1990. London.

Secretariat of the Paris Commission (1991). *Meeting of the ad hoc Working Group to Consider the Oil Content in*
    Produced Water; London 18-20 November 1991. London.

Skjaerseth, J.B. (2006). Protecting the North-East Atlantic: Enhancing Synergies by Institutional Interplay. *Marine*
    Policy, 30(2), 157.

Tan, A.K.-J. (2006). *Vessel-Source Marine Pollution; The Law and Politics of International Regulation.* Cambridge:
    Cambridge University Press.

Tatcher, M., Robson, M., Henriquez, L.R., Karman, C.C., & Payne, G. (2005). *Charm Chemical Hazard Assessment*
    and Risk Management; For the Use and Discharge of Chemicals Used Offshore: User Guide Version 1.4. Charm
    Implementation Network (CIN).

Taylor, B.G.S. (1991). Offshore Oil and Gas. *Ocean & Shoreline Management, 16,* 259-273.

Treib, O., Bähr, H., & Falkner, G. (2005). Modes of Governance: A Note towards Conceptual Clarification. *European*
    Governance Papers (EUROGOV) Retrieved December 19, 2005, from http://www.connex-network.org/eurogov/
    pdf/egp-newgov-N-05-02.pdf.

Treib, O., Bähr, H., & Falkner, G. (2007). Modes of Governance: Towards a Conceptual Clarification. *Journal of*
    European Public Policy, 14(1), 1-20.

Tromp, D., & Wieriks, K. (1994). The OSPAR Convention: 25 years of North Sea Protection. *Marine Pollution*
    Bulletin, 29(6-12), 622-626.

United Kingdom (1986). *Surveys of Chemicals Used on the United Kingdom Continental Shelf in 1982 and 1984 and the Relationship of these Chemicals to the 'notification scheme for the selection of chemicals for use offshore'; Submission to the Tenth Meeting of the Working Group on Oil Pollution; Stockholm 4-6 February 1986.* London.

United Kingdom (1988). *Concentration of Oil on Cuttings in the UK Sector; Submitted to the Twelfth Meeting of the Working Group on Oil Pollution; Oslo 9-11 February 1988.* London.

United Kingdom Offshore Operators Association (2001). *Striking a Balance; The UK Offshore Oil and Gas Industry Strategy for its Contribution to Sustainable Development 2001.* London.

Urrutia, B. (2006). The EU Regulatory Action in the Shipping Sector: A Historical Perspective. *Maritime Economics and Logistics, 8*(2), 202-221.

Van de Wateringen, S.L. (2005). *The Greening of Black Gold; Towards International Environmental Alignment in the Petroleum Industry.* Universiteit van Amsterdam, Amsterdam.

Van der Zouwen, M. (2006). *Nature Policy between Trends and Traditions: Dynamics in Nature Policy Arrangements in the Yorkshire Dales, Donana and the Veluwe.* Delft: Eburon.

Van Kersbergen, K., & Van Waarden, F. (2004). 'Governance' as a Bridge between Disciplines: Cross-disciplinary Inspiration regarding Shifts in Governance and Problems of Governability, Accountability and Legitimacy. *European Journal of Political Research, 43*(2), 143-171.

Van Oosterom, C. (2001). Hulpmiddel om M.E.R.-procedures te Versnellen? Generieke MER Olie- en Gaswinning. *Kenmerken, 8*(1), 4-7.

Van Tatenhove, J., Arts, B., & Leroy, P. (2000a). Political Modernisation. In J. Van Tatenhove, B. Arts & P. Leroy (Eds.), *Political Modernisation and the Environment; The Renewal of Environmental Policy Arrangements* (pp. 35-51). Dordrecht: Kluwer Academic Publishers.

Van Tatenhove, J., Arts, B., & Leroy, P. (2000b). *Political Modernisation and the Environment; The Renewal of Environmental Policy Arrangements.* Dordrecht: Kluwer Academic Publishers.

Vorbach, J.E. (2001). The Vital Role of Non-Flag State Actors in the Pursuit of Safer Shipping. *Ocean Development & International Law, 32*(1), 27-42.

Waters, M. (2001). *Globalization* (2nd edition). London: Routledge.

Yin, R.K. (1994). *Case Study Research; Design and Methods* (2nd edition). Thousand Oaks: SAGE Publications, Inc.

Yin, R.K. (2009). *Case Study Research; Design and Methods* (4th edition). Thousand Oaks: SAGE Publications, Inc.

Zero Discharge Group (2003). *Zero Discharges to Sea from Petroleum Activities; Status and Recommendations.*

Appendices

Appendix A: List of interviewees for the shipping case

Nr.	Institution	Date interviewed	Main subjects
1	Bunkering@Sea	10 November 2004	Bunkering
2	Royal Association of Netherlands's Shipowners	2 June 2005	Shipping sector, International Maritime Organization and European Union legislation
3	North Sea Foundation	15 June 2005	International Maritime Organization and European Union legislation
4	Ministry of Transport, Public Works and Water Management (NL)	18 July 2005a	European Union legislation
5	Ministry of Transport, Public Works and Water Management (NL)	26 August 2005b	European Union legislation and ports
6	Port of Rotterdam	31 August 2005	International Maritime Organization and European Union legislation
7	Friends of the Earth International	13 June 2007	International Maritime Organization and environmental NGOs
8	Associated British Ports*	14 June 2007	Ports
9	British Chamber of Shipping	19 June 2007	Industry self-governance
10	International Maritime Organization	19 June 2007	International Maritime Organization
11	Lloyd's Register	20 June 2007	Classification societies
12	British Ports Association	22 June 2007	Ports
13	Department for Transport (UK)	26 June 2007	International Maritime Organization
14	International Chamber of Shipping*	28 June 2007	International Maritime Organization and industry self-governance
15	Stolt-Nielsen Transportation Group	15 November 2007	Industry self-governance
16	National Ports Council and Port of Amsterdam**	22 November 2007	Ports
17	Transport&Environment	23 November 2007	Environmental NGOs
18	European Community Shipowners' Associations and European Sea Ports Organisation	23 November 2007	European Union legislation and industry self-governance
19	European Commission (DG Energy and Transport)	03 December 2007a	European Union legislation
20	European Commission (DG Environment)	03 December 2007b	European Union legislation
21	Ministry of Transport, Public Works and Water Management (NL)	25 January 2008	International Maritime Organization

* Interview with two representatives

** Interview with four representatives

Appendix B: List of interviewees for the offshore oil and gas production case

Nr.	Institution	Date interviewed	Main subjects
1	Nederlandse Aardolie Maatschappij	20 December 2005	Field development, main stakeholders, environmental effects
2	Netherlands Oil and Gas Exploration and Production Association*	05 January 2006	Environmental effects, Dutch Environmental legislation and Environmental Covenant
3	Delft University of Technology	6 March 2006	Dutch oil and gas sector
4	FO-Industry	05 April 2006	Environmental Covenant
5	Ministry of Transport, Public Works and Water Management (NL)	13 April 2006	OSPAR
6	Norwegian Oil Industry Association	13 June 2006a	Norwegian Oil Industry Association, Norway's environmental legislation
7	Hydro	14 June 2006	Industry self-governance and Norway's environmental legislation
8	Norwegian Pollution Control Authority	14 June 2006a	OSPAR
9	Norwegian Pollution Control Authority	16 June 2006b	Norway's environmental legislation and zero discharge approach
10	Bellona	16 June 2006	Environmental NGOs
11	Standards Norway*	16 June 2006	The environmental care standard of Standards Norway
12	Statoil	20 June 2006	Industry self-governance
13	Norwegian Oil Industry Association*	21 June 2006b	Norwegian Oil Industry Association, industry self-governance and Norway's environmental legislation
14	British Petroleum Norway	22 June 2006	Industry self-governance and Norway's environmental legislation
15	Norwegian Petroleum Directorate*	22 June 2006	Norway's environmental legislation and zero discharge approach
16	Institute for Marine Resources & Ecosystem Studies	15 September 2006	Industry self-governance
17	State Supervision of Mines (NL)	18 September 2006	Environmental Covenant and OSPAR
18	Wintershall NL	25 September 2006	Industry self-governance and Environmental Covenant
19	Department of Trade and Industry (UK)	05 October 2006a	Oil and Gas Industry Task Force
20	Policy Studies Institute (UK)	06 October 2006	Produced water research project and UK oil and gas sector
21	OSPAR Secretariat	09 October 2006	OSPAR

Appendix B: Continued.

Nr.	Institution	Date interviewed	Main subjects
22	International Association of Oil & Gas producers	13 October 2006	International Association of Oil & Gas Producers and industry self-governance
23	British Petroleum UK	16 October 2006	Industry self-governance and UK's environmental legislation
24	Department of Trade and Industry (UK)	17 October 2006b	UK's environmental legislation
25	Department of Trade and Industry (UK)	19 October 2006c	OSPAR and UK's environmental legislation
26	Consultant Offshore Industry (UK)	20 October 2006	Oil and Gas Industry Task Force and industry self-governance
27	European Oilfield Speciality Chemicals Association	23 October 2006	European Oilfield Speciality Chemicals Association and OSPAR
28	United Kingdom Offshore Operators Association	23 October 2006	United Kingdom Offshore Operators Association and industry self-governance
29	Ministry of Housing, Spatial Planning and the Environment (NL)	23 November 2006	Situation before and development of Environmental Covenant
30	GazdeFrance NL	10 January 2007	Industry self-governance and Environmental Covenant

* Interview with two representatives

Appendix C: Direct observation; meetings and seminars visited

Meeting	Date	Case
Green Port Conference Rotterdam	4-5 March 2005	Shipping
Sustainable Shipping Conference	17-18 May 2005	Shipping
Meeting 'Nationale werkgroep' for the Marine Environmental Protection Committee 53 (preparation of the Dutch delegation for MEPC 53)	19 June 2005	Shipping
Bek&Verburg (port reception facility Rotterdam Harbour)	20 September 2005	Shipping
Meeting of 'Werkgroep Chemicaliën' (working group that is part of the Dutch Environmental Covenant)	28 April 2006	Offshore oil and gas production
North Sea Ministerial Conference	3-5 May 2006	Shipping and offshore oil and gas production
Meeting 'Nationale werkgroep' for the Marine Environmental Protection Committee 55 (preparation of the Dutch delegation for MEPC 55)	19 September 2006	Shipping
Marine Environmental Protection Committee Meeting 55	9-13 October 2006	Shipping
Meeting 'Klankbordgroep Havenontvangstinstallaties' (working group on port reception facilities)	22 November 2007	Shipping
Seminar 'Opportunities for Clean Shipping'	12 March 2008	Shipping

Summary

The protection of the marine environment is a relatively new area of study within governance literature. Yet, the long history of environmental governance at sea and the special characteristics of the sea make the protection of the marine environment an interesting policy domain for a study into how environmental governance practices and the authority of the state have changed under influence of globalization in the last decades. The scale and intensity of maritime activities is growing, increasing not only the call for effective environmental governance, but also increasing the stakeholders and stakes involved in the governance of the marine environment. State sovereignty is decreasing the further one gets away from the coastline, which means that states do not always have the autonomy to develop policies. The marine environment is a transboundary area where stakeholders with different interests have to cooperate to protect the marine environment from the adverse effects of the intense use of the sea. That is also why transnational cooperation in the environmental governance of the sea is – in fact – not that new; the environmental effects of shipping and offshore oil and gas production in the North Sea are already regulated since respectively the 1950s and the late 1970s. The objective of this research is therefore *to analyse how shifts in governance affect and change the authority of the state in the environmental governance of shipping and of offshore oil and gas production in order to generate new insights for the governance debate.*

Conceptual framework

The shift from conventional governance to new, innovative governance arrangements is based on four shifts in governance; a shift from single to multiple actors, multiple levels, multiple rules of the game and multiple steering mechanisms. These four shifts in governance and the plethora of governance arrangements that currently exist present social scientist with new theoretical challenges. Conventional theoretical approaches are insufficient to understand the shifts in governance, because they generally suffer from methodological nationalism. New theoretical concepts are therefore needed to understand the emergence of shifts in governance, the effect of these shifts onto existing and new governance arrangements and the changing authority of the state. That is why a new conceptual framework is developed to analyse how shifts in governance affect and change the authority of the state in the environmental governance of shipping and of offshore oil and gas production

This conceptual framework is based on the concept sphere of authority that is developed by James N. Rosenau and the policy arrangement approach as developed by Van Tatenhove and colleagues. In this thesis, the concept sphere of authority is defined as *the temporary stabilization of the organization and substance of a policy domain within which actors take decisions through the development of steering mechanisms and/or have the ability to generate compliance with these steering mechanisms.* A sphere of authority is characterized by its organization, substance and results. The element of organization consists of the three dimensions actors, power resources and relations, and rules of the game. The substance of a sphere of authority is defined by the dimension discourse, i.e. those interpretative frameworks or dominant interpretative schemes, which give understanding and meaning to a policy domain. The results of a sphere of authority consist of two dimension: steering mechanisms (the instruments through which change is achieved) and compliance mechanisms (the instruments through which compliance is achieved).

Besides the sphere of authority concept, this thesis also defines two mechanisms through which changes in governance and authority come about. The first mechanism of policy change is the renewal of an existing sphere of authority. Renewal takes place when the interaction between actors and the results in a sphere of authority changes. Renewal starts when a dimension of a sphere of authority changes and causes a chain reaction throughout the whole sphere of authority. The second mechanism of policy change is the emergence of a new sphere of authority. A new sphere of authority emerges when a new collectivity of actors interacts to develop steering and compliance mechanisms.

Research questions and methods

Based on the conceptual framework the following research questions were defined for this thesis:

- Which policy changes have taken place since the emergence of the first sphere of authority in the environmental governance of shipping and of offshore oil and gas production?
- What is the effect of these policy changes in terms of the renewal of existing and the emergence of new spheres of authority in the environmental governance of shipping and of offshore oil and gas production?
- To what extent has the renewal of existing and the emergence of new spheres of authority led to shifts in governance in the environmental governance of shipping and of offshore oil and gas production?
- How has the authority of the state in developing steering mechanisms and generating compliance changed as a result of the policy changes in the environmental governance of shipping and of offshore oil and gas production?
- What are insights gained from the development and application of the theoretical concepts of this thesis and the results of the empirical analysis for the governance debate?

This thesis is based on case study research. The most important source of data were semi-structured interviews; 21 interviews were done for the shipping case and 30 for the offshore oil and gas production case. For the shipping case these interviews were done in the Netherlands and the UK, while in the offshore oil and gas case respondents from the Netherlands, Norway and the UK were selected. For both cases, governmental officials, industry actors (mainly industry associations) and environmental NGOs have been interviewed. Besides interviews, documents from the most important meetings of the regulating bodies within the environmental governance of shipping and offshore oil and gas production were gathered. The documents served as support for the information derived from the interviews. The last and third method of data collection has been direct observation at meetings and seminars where stakeholders within the environmental governance of shipping and offshore oil and gas production meet each other.

Changing environmental governance of shipping and offshore oil and gas production

The traditional spheres of authority in the environmental governance of both shipping and offshore oil and gas production were remarkably similar. In both cases, the traditional spheres of authority evolved around a transnational forum that developed steering mechanisms, i.e. respectively the International Maritime Organization and the Paris Convention. The difference is that the IMO sphere of authority emerged in the 1950s and the Paris sphere of authority in the 1970s. Both state and industry actors were involved, but structural involvement of environmental NGOs was lacking. Both transnational forums developed transnational steering mechanisms. The IMO in the form of international conventions and the Paris Convention in the form of decisions and recommendations. Yet, compliance with those transnational steering mechanism relied mostly on formal, national compliance mechanisms.

Both cases have been subject to two policy changes. However, the nature of the policy changes differed, resulting in differences in the development of the environmental governance of shipping and of offshore oil and gas production. In shipping, a policy change took place when the UN Convention on the Law of the Sea was adopted in 1982. UNCLOS provided new rules of the game for the IMO sphere of authority, because port states gained the right to develop steering and compliance mechanisms for shipping, something that had been the exclusive competence of flag states until then. The second policy change is the emergence of the EU as a rival sphere of authority in specific environmental issues between 1995 and the present. Because of two tanker incidents in 1999 and 2002, the EU has been developing European steering and compliance mechanisms. However, within the IMO sphere of authority, the discourse of protecting the level playing field for the shipping industry by developing global standards is dominant. Since this discourse is challenged by the development of EU steering and compliance mechanisms, the IMO is far from enthusiastic about the EU sphere of authority. It is important to note that the EU sphere of authority only emergences in those cases where the EU (threats to) develop(s) EU steering mechanisms.

The first policy change in the environmental governance of offshore oil and gas production has been the emergence of new spheres of authority. In the 1990s, national spheres of authority emerged in the UK, Norway and the Netherlands, while at the same time the industry developed a system of self-governance through industry steering and compliance mechanisms. As a result, all three national spheres of authority developed steering mechanisms that are characterized by joint initiatives between governmental actors and the industry. The second policy change is, similar to shipping, the emergence of an EU sphere of authority. The EU does not have a clear policy strategy for regulating the offshore industry, but has been regulating the industry by including them in directives focused on a variety of issues and sectors. While the EU sphere of authority have had a considerable impact on the national spheres of authority, the EU sphere of authority has not led to a renewal of the OSPAR sphere of authority. The national spheres of authority have renewed, because they have implemented the EU steering mechanisms through national law, which has partly undermined the cooperation between the government and the industry in the national, joint steering mechanisms.

The policy changes and their effects on the environmental governance of shipping and of offshore oil and gas production have led to several shifts in governance. The first shift, from single to multiple actors has affected both cases; the policy changes have led to more structural involvement of industrial actors and environmental NGOs. The shift to multiple levels has also occurred in both cases, despite the fact that multiple levels have always been a core feature of the environmental governance of both maritime activities. The environmental governance of shipping is only in a limited way affected by the third shift in governance; a shift to multiple rules. Traditionally, the environmental governance of shipping has been mostly embedded in formal institutions. Only with the emergence of the new EU dynamics, informal rules of the game (i.e. about how to deal with EU within the IMO) have emerged. In contrast, in the environmental governance of offshore oil and gas production the shift to multiple rules has been more substantial, because of the emergence of new spheres of authority. With regard to the shift to multiple steering mechanisms, shipping has only known a very limited shift from IMO steering mechanisms to some EU steering mechanisms. This shift to multiple steering mechanisms has been much more important in the environmental governance of offshore oil and gas production where besides the Paris/OSPAR steering mechanisms, European, national and industry steering mechanisms have emerged.

Changing authority of the state

Traditionally, state actors within the transnational forums had the formal authority to develop steering mechanisms. Within shipping these were the flag states, while in offshore oil and gas production, it were those states with platforms in their territorial waters or exclusive economic zones. Yet, the industry actors had authority as well, because they introduced best practices and techniques which often formed the basis for the development of new standards. In fact, this input and expertise was explicitly sought by the states.

Because of the policy changes in the environmental governance of shipping, one of the main changes in the authority of the state is that formal authority has become shared between flag and port states. In addition, the power struggle over authority became one between flag, port and coastal states, environmental NGOs and industry actors. In this power struggle, the authority of flag and port states is based on the formal competence to develop steering mechanisms, while the power of coastal and environmental NGOs is back-up by the EU port state coalition which strives for more stringent standards. The power of the industry is based on its expertise and best practices. In the last decade, the environmental coalition and the port states seem to be most effective in this power struggle, since standards that go beyond existing technologies have been adopted.

In the environmental governance of offshore oil and gas production, the authority of the state in the development of steering mechanisms has become more complex because of the multiple spheres of authority that govern the offshore industry. The authority of the state increased, because they not only develop OSPAR steering mechanisms, but also national ones. Yet, nationally, the authority to develop steering mechanisms is shared between governmental and industry actors. On top of that, the industry gained authority by developing industry steering mechanisms. The interplay between the industry self-governance, the regional and national spheres of authority means that authority is increasingly shared between the state actors and the industry. At the same time, the EU sphere of authority has led to a loss of authority of both state actors that are directly involved in regulating offshore oil and gas production and the offshore industry itself, because these actors are not at the heart of the EU sphere of authority.

Traditionally in both shipping and offshore oil and gas production, the (flag) state was responsible for implementation and arranging ways to ensure compliance. In practice, however, this authority of the state was more complex, because even though states were responsible, it is the industry that in the end had to live up to the standards and effective compliance mechanisms in especially the shipping case were missing.

In the environmental governance of shipping, the way in which compliance is assured changed considerably as a result of the policy changes that have taken place. First, port states have gained rights to develop steering mechanisms and they have taken this opportunity to develop a system of Port State Control. Second, flag states have developed compliance mechanisms as well. Besides this increased authority of both port and flag states, the European Commission has a formal compliance mechanism with the infringement procedure and the IMO has developed voluntary compliance mechanisms. All in all, the authority of the state in generating compliance has increased at the expense of the authority of the industry. Moreover, there are other actors that have gained some authority over compliance.

After the policy changes, the authority over generating compliance in the environmental governance of offshore oil and gas production has become shared between the state and the industry. Both have increased the use of compliance mechanisms. Moreover, the state and the industry are interdependent when it comes to generating compliance, because the industry monitors its emissions and discharges and because of safety reasons inspections on platforms can only be done when announced beforehand.

Conclusions and reflection

An important new insight that this thesis generates for the governance debate is that changing governance practices and changing authority in governance should be studied in a broader context. First, because empirical results on new spheres of authority are usually compared with a theoretical narrative about conventional spheres of authority. Yet, this research shows that an empirical account of the traditional sphere of authority can differ from the theoretical construct of conventional governance practices. Second, despite the fact that most of the changes in governance come about through the development of new spheres of authority, one should take into account that new governance arrangements co-exist with already existing governance arrangements. Changing practices and authority are only fully understood when one looks at new spheres of authority and the effect of new spheres of authority on the traditional one(s).

A second insight that this thesis generates for the governance literature is that the diffusion of authority over different actors was confirmed in this thesis. However, this thesis also shows that diffusion of authority is not the same as the zero-sum game of less authority for states and more for non-state actors. The actors through which authority diffuses also do not have to be private actors, but can be public actors as well. In this regard, it can also be concluded that the diffusion of authority to non-state actors is more likely in new spheres of authority, where the breeding ground for innovation is better than in existing spheres of authority.

In reflecting on this thesis, it was concluded that it was sometimes difficult to match the objective of analysing processes of change in governance from a long-term perspective with the objective of studying the implications for the authority of the state in more detail. In reflecting on the conceptual framework, it is concluded that the concept sphere of authority is broad enough to cover various types of governance arrangements, including conventional ones. The concept can also be used to analyse the dynamics within a sphere of authority and the resulting authority of actors in developing steering mechanisms and generating compliance. Yet, the true merits of this conceptual model in linking the dynamics within a sphere of authority with the resulting authority of actors can only be further assessed when in-depth studies are done. Finally, the concepts policy change and renewal of a sphere of authority help in generating insight into processes of change in governance. However, this thesis also showed the difficulty in pinpointing when something can be regarded as a policy change or as the renewal of a sphere of authority and when not.

About the author

Judith van Leeuwen was born in Wilnis (the Netherlands) on 3 August 1980. After finishing her secondary education at the Eckart College in Eindhoven in July 1998, Judith moved to Nijmegen to study Political Science of the Environment at the *Katholieke Universiteit Nijmegen* (the former Radboud University Nijmegen). She graduated as an MA from this program in September 2003. A year later, in July 2004, Judith also obtained the MSc diploma in Political Sciences from Radboud University Nijmegen.

Judith continued her academic career at the Environmental Policy Group of Wageningen University from October 2004 to the present. At this group, she started with her NWO-funded PhD-project entitled 'Governance and gradations in traditional sovereignty: the case of off shore activities in the North Sea area' in October 2004. This project ended in December 2008. Since January 2009, Judith is lecturer at the Environmental Policy Group and study advisor of the BSc and MSc Environmental Sciences and the MSc Urban Environmental Management. Since January 2010, she is also researcher at the Environmental Policy Group.

Printed in the United States
by Baker & Taylor Publisher Services